Undergraduate Lecture Notes in Physics

Undergraduate Lecture Notes in Physics (ULNP) publishes authoritative texts covering topics throughout pure and applied physics. Each title in the series is suitable as a basis for undergraduate instruction, typically containing practice problems, worked examples, chapter summaries, and suggestions for further reading.

ULNP titles must provide at least one of the following:

- An exceptionally clear and concise treatment of a standard undergraduate subject.
- A solid undergraduate-level introduction to a graduate, advanced, or non-standard subject.
- A novel perspective or an unusual approach to teaching a subject.

ULNP especially encourages new, original, and idiosyncratic approaches to physics teaching at the undergraduate level.

The purpose of ULNP is to provide intriguing, absorbing books that will continue to be the reader's preferred reference throughout their academic career.

Series editors

Neil Ashby
Professor Emeritus, University of Colorado, Boulder, CO, USA

William Brantley
Professor, Furman University, Greenville, SC, USA

Matthew Deady
Professor, Bard College Physics Program, Annandale-on-Hudson, NY, USA

Michael Fowler
Professor, University of Virginia, Charlottesville, VA, USA

Morten Hjorth-Jensen
Professor, University of Oslo, Oslo, Norway

Michael Inglis
Professor, SUNY Suffolk County Community College, Long Island, NY, USA

Heinz Klose
Professor Emeritus, Humboldt University Berlin, Berlin, Germany

Helmy Sherif
Professor, University of Alberta, Edmonton, AB, Canada

More information about this series at http://www.springer.com/series/8917

Ross Barrett · Pier Paolo Delsanto
Angelo Tartaglia

Physics: The Ultimate Adventure

 Springer

Ross Barrett
Rose Park, SA
Australia

Pier Paolo Delsanto
Dipartimento di Scienza Applicata e
  Tecnologia (DISAT)
Politecnico di Torino
Turin
Italy

Angelo Tartaglia
Dipartimento di Scienza Applicata e
  Tecnologia (DISAT)
Politecnico di Torino
Turin
Italy

ISSN 2192-4791   ISSN 2192-4805   (electronic)
Undergraduate Lecture Notes in Physics
ISBN 978-3-319-81097-3   ISBN 978-3-319-31691-8   (eBook)
DOI 10.1007/978-3-319-31691-8

I became a physicist to understand the world,
then I became a writer to try and change it

Carla H. Krueger

Foreword

Early last century, as revolutionary painters George Braque and Pablo Picasso in Paris developed a new vision of painting that would soon be dubbed 'Cubism', patent clerk Albert Einstein in Bern, Switzerland, was developing new theories that launched the domain of physics into another universe. Or rather, he catapulted our understanding of the universe into a new dimension. Just as Braque and Picasso were inspired by and extended monumental ideas first proposed by Paul Cézanne, Einstein constructed his great advances on the foundations built by the generations of physicists who went before him.

If you understand the space–time continuum or have no idea what that means, this book is for you. If you've ever wondered why Sir Isaac Newton is a giant among scientists, this book is for you. If you are interested in how things work in the physical world and how the physical sciences developed, this book is for you. Edwin Herbert Land, American scientist and inventor of the Land camera, stated, "Don't do anything that someone else can do. Don't undertake a project unless it is manifestly important and nearly impossible." That is precisely what the authors have undertaken. The beauty of this book is that it comprises essentially all of physics described in language comprehensible to the non-scientist. The authors present difficult concepts, which would normally be accompanied by pages and pages of mathematics, in lucid English with clear straightforward figures. And they have accomplished this feat while making it a good read. In short, this book is for every curious person.

The authors are a group of individuals each with a broad background in the physical sciences. Furthermore, their collective experience and knowledge embraces the arts as well as science. I have listened with pleasure to Pier-Paolo Delsanto, a Senior Professor from the Politecnico of Turin, Italy recite from Horace in the original Latin. His passion for the beauty of Latin is equal to his passion for the beauty of physics. Physicist Ross Barrett, former Research Leader at the Defence Science and Technology Organisation, Adelaide writes plays for the live stage, many of which have been produced in professional partnerships in Australia. Angelo Tartaglia, Senior Professor from the Politecnico of Turin is the author of a

theory identifying dark energy with the strain energy of a four-dimensional continuum that accounts for the accelerated expansion of the universe. If you don't
understand what that means, read this book and you will!

I enthusiastically urge you to read this book. Skip the parts you don't understand. Read on. Discover the passion and beauty of physics and understand how it
affects your life every day and in surprising ways.

> Heat cannot be separated from fire
> Or beauty from the eternal
>
> —Dante Alighieri

Paul Allan Johnson
Senior Physicist at the Los Alamos
National Laboratory and Artist

Preface

Just imagine this scene: a physicist at a party mentions his/her profession casually to a new acquaintance. In most cases the reaction is a puzzled look and protests such as "But Physics is so dry!" or "I could never understand it" or "At school, Physics was my *bête noire*."

We believe that physics, far from being dry, can be and should be made beautiful, inspiring and enjoyable. For many students or casual readers, physics may indeed be hard, but the difficulty stems usually from the mathematical formalism which is used to explain it. Even a children's story can be extremely hard to understand, if it is narrated in a language unknown to the listener. Mathematics is the language of physics. It is requisite, if the goal is scientific research or nontrivial applications. It may even be your best guide in subfields, such as atomic and nuclear physics, where many of the concepts and results are almost in contradiction with our daily experience, and the abstractions of quantum mechanics prevail.

Yet, a basic understanding of the achievements of modern physics should be part of the culture of each of us, just as well as a basic knowledge of music, literature and art. Giants, such as Einstein, Bohr, Heisenberg and Gell-Mann (to quote just a few), represent pinnacles of human creativity and ingenuity, just as well as Shakespeare, Leonardo, Beethoven and Bach. Everybody should have access to the wonders and glamour of modern physics, even if only a few possess the mathematical tools, which are usually required for a deeper understanding.

Thus the goal of our book is to simplify the path to those who have the intellectual curiosity, but not the mathematical skills, which are needed to approach physics through the customary channels. But at the same time we need to stress that we wish to simplify, but *not* oversimplify. As in the famous aphorism attributed to Einstein: *everything should be made as simple as possible, but not simpler*. Our goal is divulgation, yet we wish to maintain a solid scientific style. This is necessary, because physics is *not* a fairy tale from some imaginary world, even if sometimes its abstract nature makes it appear as such. We must learn to distinguish bad physics from good physics and to understand, when we read an article in a newspaper what is the likely truth behind the patronizing words of the journalist.

Our journey begins in the first three chapters with an introduction to what physics is and what it is not (or should not be). We also provide some of the essential mathematics, kept as elementary as possible, and a glimpse of the world of experimental physics. Physics is, after all, an *a posteriori* science, i.e. it must begin from the observation of the phenomenology around us.

Then, in the next three chapters, we continue with what is currently known as *Classical Physics*. Some of the readers will probably be more interested in *Modern Physics*, i.e. in the developments of physics from the beginning of the twentieth century, including relativity and quantum mechanics. However, Newton's intuition about gravity (i.e. that it is the same force on the surface of the earth and among celestial bodies) and Maxwell's unification of electricity and magnetism are truly awe-inspiring. Nowadays they are so much ingrained in our cultural environment, that they seem almost obvious. Toddlers, who keep dropping all kinds of things from high chairs to the despair of their parents, are better physicists than their elders because they have not yet lost their sense of wonder.

The next five chapters are devoted to modern physics. Relativity and quantum mechanics are first introduced and then applied to the study of the extremely small (Atomic and Nuclear physics, Elementary Particles) and of the extremely large (the Universe itself). The following Chap. 12 is the odd man out, since it abandons mainstream physics to follow the ever expanding field of the application of physics methodologies to multidisciplinary problems.

Finally, Chap. 13 tries to present a foretaste of the future, starting with a discussion of current *open problems*. In order to reverse the established fact that *short-term scientific predictions are always too optimistic, while long-term predictions are invariably too timid*, this chapter allows some amusing speculations, belonging maybe more to the realm of science fiction than of physics, but rigorous in preserving logical consistency.

To conclude, the goals of this book are a continuous quest for simplicity within the constraints of scientific accuracy over a broad range of modern physics. Consequently, in our opinion, it provides useful background reading and tools for those who would like to study physics, even if not as their main discipline. For those who are beginning a physics degree, it provides an overview of the entire subject, before they immerse themselves in the technical details of some of its many specialized branches. Finally, it will hopefully answer some of the questions that tantalize the *armchair philosopher*, who resides in all of us at all ages.

Acknowledgement

It is our pleasure to thank Dr. Matteo Luca Ruggiero for his valuable and most efficient help in finalizing our draft.

Ross Barrett
Pier Paolo Delsanto
Angelo Tartaglia

Contents

Chapter 1
The Whats and Wherefores of Physics

Physics is the ultimate intellectual adventure, the quest to understand the deepest mysteries of our Universe. Physics doesn't take something fascinating and make it boring. Rather, it helps us see more clearly, adding to the beauty and wonder of the world around us. When I bike to work in the fall, I see beauty in the trees tinged with red, orange and gold. But seeing these trees through the lens of physics reveals even more beauty.

Tegmark [1]

Abstract In this chapter we set out the scope of the book and the relationship of physics to other scientific disciplines and philosophy. We also briefly discuss the partition into Classical and Modern Physics, the practical relevance of the topic and some of the a priori tools guiding physicists in their endeavour. In order to streamline the information, particularly for readers more interested in the recent developments in the field, in this Chapter and in the following one, much of the information has been relegated to self-explanatory figures and tables.

1.1 The Beginning

In the beginning was Philosophy, and Philosophy was with Science, and Science was Philosophy. Philosophy shineth in darkness; and the darkness comprehended it not.

But then Philosophy begat children. And the children were many and very successful.

Already in the 6th Century the philosopher Boethius mentioned the three liberal arts of the Trivium,[1] i.e. grammar, logic and rhetoric, and the four arts of the Quadrivium,[2] i.e. arithmetic, geometry, music and astronomy. In addition, there were the practical arts, such as medicine and architecture. Later came the division into subjects or disciplines as we know them today, such as physics, chemistry,

---

[1]In Latin, a place where three roads meet.
[2]In Latin, a place where four roads meet.

© Springer International Publishing Switzerland 2016
R. Barrett et al., *Physics: The Ultimate Adventure*, Undergraduate Lecture
Notes in Physics, DOI 10.1007/978-3-319-31691-8_1

1

biology, psychology, etc. In the same manner as a most generous mother, Philosophy divested herself of most of her possessions in favour of her children.

Humans, being human, attribute to themselves the mission of trying to understand the nature of the wonderful and complex world in which they happen to live. For such a quest they utilize epistemological tools, which can be either a priori, i.e. within oneself, or a posteriori, i.e. outside oneself. For the sake of an easy classification, we call mathematics the former and science the latter. Mathematics itself is not a science, yet it constitutes an invaluable tool for science, since it provides the means to interpret the reality we observe and a methodology to search for the underlying governing laws (assuming as a working conjecture that such laws exist and are immutable).

1.2   What Is Physics?

Among all scientific disciplines, physics is ideally suited to the task of discovering the laws of nature, since it deals with everything from the smallest particles (electrons or neutrinos), to the largest known entities (galaxies, clusters of galaxies, or even the universe itself): see Fig. 1.1. Physics also encompasses an astonishing range of times: see Fig. 1.2. There is, however, a remarkable region at sizes around 10^{-8} m, i.e. at the molecular level, which is predominantly within the domain of chemistry. This does not mean that molecular physics is not relevant for physicists. It is, but at that level, chemical reactions, which are extremely important both from an applicative and a theoretical point of view, are prevalent and they can best be studied with a very different methodology (that of chemistry). In the last few decades, however, the development of applied Quantum Mechanics has allowed the two converging disciplines of physical chemistry and chemical physics to emerge.

A subfield of chemistry, organic chemistry, has its own basic relevance, since it is a prerequisite to the understanding of life, i.e., to biology (although not to an understanding of how life originated, which still remains a very open question). Life itself appears in a wide variety of forms. In fact an estimated number of 10 million species fill the realms of fauna and flora (in addition to more primitive forms of life), spanning about eight orders of magnitude in their linear dimensions and 24 in their masses (see Fig. 1.1).

To appreciate the range of our task in this book, it may be helpful to look at how the field of physics is subdivided (see Fig. 1.3). A *caveat*, however, is needed, since the various branches are interlaced by means of a thick and ever expanding network of links, which are by necessity omitted in the figure. In addition, *Universality*, a conjectured, but not yet well investigated property of the physical world, foretells the emergence of other similarities and analogies among apparently unrelated phenomenologies (see Chap. 12).

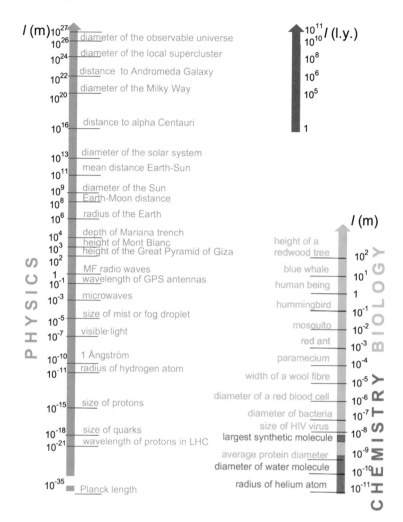

Fig. 1.1 The spatial domain of Physics from the so-called *Planck length* 1.616252×10^{-35} m, below which the concept itself of dimension loses any physical meaning, up to the current estimate of the diameter of the Universe (approximately 13.8×10^9 l.y.). Selected lengths or distances are reported in logarithmic scale, encompassing more than 60 orders of magnitude. For astronomical distances a more suitable scale in light-years is also reported (1 l.y. $= 9.4605284 \times 10^{15}$ m). On the right side selected lengths of relevance in Chemistry and Biology are included

Although our goal is declaredly *not* the history of physics, but rather physics itself, we report in Fig. 1.4a, b a concise chronology of some of the most relevant advances, mainly since the advent of classical physics, which we can date to the time of Galileo. This is not to be taken that no physics, nor even no relevant

Fig. 1.2 The temporal domain of Physics from the *Planck time*, i.e. the time required for light to travel in vacuo the distance of one *Planck length* to the currently estimated age of the Universe (13.8×10^9 years, to be compared with estimates of around 7000 years by ancient biblical scholars). Selected durations in seconds (to the right also in years for longer lasting events) in a logarithmic scale, encompassing more than 60 orders of magnitude

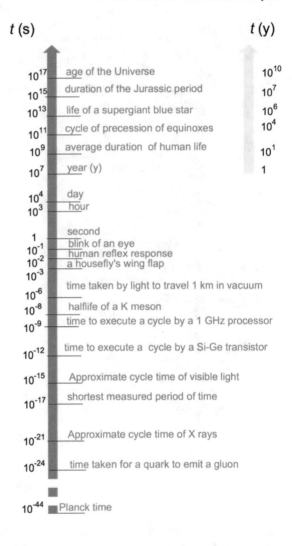

physics, was done before Galileo. As we mentioned before, ancient philosophers were also physicists at heart in their curiosity about nature, and many of them left important contributions, particularly in astronomy. However, Galileo was the first to introduce in a systematic way the scientific methodology of modern science, in opposition to the *Ipse dixit* (*He said it, so it is so*) attitude of some previous thinkers, and to pure logical deduction, without reference to facts.

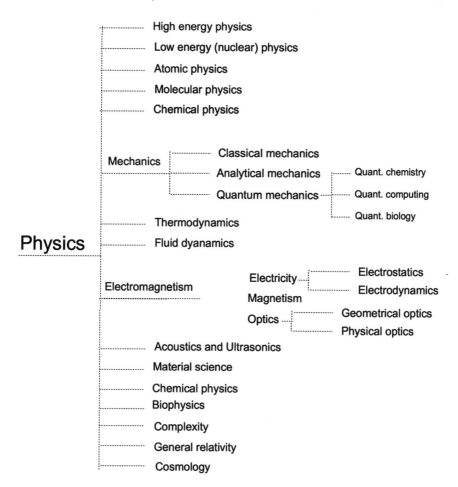

Fig. 1.3 Traditional subfields of Physics. However, due to the ever growing cross-fertilisation among different methodologies and applications, the boundaries among subfields and even between Physics and other disciplines tend to become more and more fuzzy and arbitrary. Also new subfields (or specialties) are continuously being born

1.3   Classical and Modern Physics

From Fig. 1.4a, b we may see that physics can be divided into *classical* and *modern physics*, the former including basically all the physics that was known at the end of the 19th century. Nowadays classical physics has perhaps become less glamorous than its modern counterpart, but nevertheless it must be included in our book (Chaps. 4–6), since it is a necessary prerequisite to the understanding and appreciation of modern physics. A remarkable distinction between the two is that classical physics encompasses, at least at first glance, all the phenomenology within the

(a)

- 5th century B.C. Leucippus, Democritus: atomism
- 4th century B.C. Aristotelian Physics
- 240 B.C. Eratosthenes: Estimate of the circumference of the Earth
- 3th century B.C. Aristarchus of Samos: heliocentric model
- 3th century B.C. Archimedes
- 150 A.D. Ptolemy publishes Almagest
- ~1000. Ibn al-Haytham: Theory of Optics.
- ~1100. Omar Khayyám: Calculation of the Solar Year
- 1512. Nicholas Copernicus: heliocentric theory
- 1577. Tycho Brahe: proves that comets are distant object and not athmosferic phenomena
- 1589. Galileo Galilei: experiments on inertia
- 1609-1619. Johannes Kepler: Laws on planetary motions
- 1621. Willebrord Snell: law of reflection
- 1665. Isaac Newton: law of universal gravitation
- 1675. Ole Rømer: first estimate of the speed of light
- 1678. Christiaan Huygens: principle of wavefront sources
- 1687. Isaac Newton publishes his "Principia"
- 1705. Edmond Halley predicts the periodicity of Halley's comet
- 1738. Daniel Bernoulli publishes "Hydrodynamica"
- 1785. Charles Coulomb: inverse-square law of electrostatics
- 1789. Antoine Lavoisier: conservation of mass
- 1798. Henry Cavendish measures the mass of the Earth and the Gravitational constant
- 1800. Alessandro Volta invents the electric battery
- 1820. Hans Oersted proves that a current deflects a compass needle
- 1820. André-Marie Ampère quantifies the force between parallel currents
- 1821. Michael Faraday builds an electricity-powered engine
- 1824. Sadi Carnot: works on heat engines
- 1826. Georg Ohm: law of electrical resistence
- 1827. Robert Brown discovers Brownian motion

Fig. 1.4 a Chronology of some of the most relevant discoveries and achievements in Physics (Part I). **b** Chronology of some of the most relevant discoveries and achievements in Physics (Part II). The selection is, of course, arbitrary and often unfair, since in many cases a discovery is just the end result of the work of many previous researchers. Also for many scientists, e.g. for those living outside the Western world, it may be very hard to have their work published and/or acknowledged

(b)
- 1831. Michael Faraday discovers the electromagnetic induction
- 1849. James Prescott Joule shows that heat is a form of energy
- 1865. James Clerck Maxwell publishes his works on electromagnetism. Later on, he realizes that light is electromagnetic radiation.
- 1874. William Thomson (Lord Kelvin): second law of thermodynamics
- 1888. Heinrich Hertz discovers radio waves
- 1895. William Röntgen discovers X-rays
- 1896. Antoine Becquerel discovers the radioactivity of uranium
- 1897. Joseph Thomson discovers the electron
- 1900. Max Plank states his radiation law.
- 1905. Albert Einstein publishes his theory of special relativity
- 1913. Niels Bohr presents his quantum model of the atom
- 1913. Robert Millikan measures the elementary charge (e.g. of an electron)
- 1915. Albert Einstein publishes his theory of general relativity
- 1924-27. Werner Heisenberg, P.A.M. Dirac, Erwin Schrödinger, Wolfgang Pauli, Max Born and others develop quantum mechanics
- 1929. Edwin Hubble discovers the expansion of the Universe
- 1932. James Chadwick discovers the neutron
- 1932. Carl Anderson discovers the positron
- ~1950. Richard Feynman, Julian Schwinger and Sin-Itiro Tomonaga develop Quantum Electrodynamics
- 1958. Charles Townes invents the laser
- 1963. Murray Gell-Mann and George Zweig propose the quark model
- 1965. Arno Penzias and Robert Wilson discover the 3K background radiation
- 1960s. Sheldon Glashow, Steven Weinberg, Abdus Salam develop the Standard Model
- 1998. Observation ot the accelerated expansion of the Universe
- 2012. The Higgs boson is found and (to a very high level of confidence) identified

Fig. 1.4 (continued)

range of our common experience, while modern physics concerns objects that are either too small (molecules, atoms, atomic nuclei and the so-called elementary particles) or too fast (moving at, or close to the speed of light) to play much part in our ordinary daily lives. The physics of these objects defies our everyday experience, so that we cannot really *explain* it in the usual sense of the word, i.e. "*to make plain, manifest or intelligible*" (e.g. by reducing the new concepts to an elaboration of old ones).

Thus the only way to understand quantum mechanics or relativity, which form the framework for almost all modern physics, is through the Arianna's thread of mathematics. As the Italian poet Dante Alighieri wrote:

You sailors in your little boats that trail
My singing ship because so keen to hear,
By now it might be time for you to sail
Back till you see your shoreline reappear,
For here the sea is deep, and if you lose
My leading light just once, then steering clear
Might bring bewilderment... [2]

In order to allow the reader *not* to lose our *leading light*, we will need some mathematical tools, which we will provide as needed and will strive to keep to an absolute minimum, by attempting as far as possible to avoid formulas.

1.4   Why Do We Need Physics?

Before we continue further, it may be wise to ask ourselves whether we really need physics. In fact, the common perception of science by laymen varies wildly from uncritical acceptance to disbelief, often, curiously enough, in the same person and within a short period of time. A typical reaction of somebody, whose ingrained beliefs are challenged by some scientific fact, is to state polemically that "*also scientists may be wrong*". While this is undoubtedly true, it is a fact that we depend for almost all we do every day on technologies that have been developed thanks to science. Everything from the simple tungsten lamp to PC's, cellular phones, satellite navigators, etc. provide a most compelling proof of the validity of the laws of physics on which they are based. If science has changed our way of life so dramatically, it should also be expected to change our way of thinking. The question is therefore not whether we should believe in science, but rather in which science to believe, i.e. how to recognize good from bad science, and also to realize that even good science has its limits of validity. This will be the subject of Chap. 3.

Besides Relativity and Quantum Mechanics, another recent branch of modern physics, which also yields unexpected and counter-intuitive results is *Complexity* (see Fig. 1.3). Here the novelty is due not to the size or speed of the objects, but to their extremely large number, which in itself is rather surprising. In fact it is normally easier to treat a *many body system* than one comprising only a few bodies. Even the relative motion of only three, non-trivially interacting bodies cannot be predicted analytically (which is perhaps as surprising as the demonstrated non-solubility of algebraic equations of the fifth order or higher). If the number of objects is large, but not too large (say thousands or millions, depending on the computational facilities), even numerical solutions by means of large scale computers become very time consuming and not too reliable. Large systems (e.g. in the field of economics) can be studied with more ease by means of statistical techniques but, as we will see in Chap. 12, not when *complexity* occurs.

1.5 Beauty and Symmetries

To conclude, it may be useful to spend a few words on the tools of physics. However, since many treatises on the philosophy of science can be found, both from the point of view of a philosopher and of a scientist, we prefer to mention a much less debated, but nevertheless very powerful source of inspiration which often guides a scientist in his/her endeavour, namely *beauty*. Scientists are, often unconsciously, moved by beauty, just as was Lucretius, who in the proem to his *"De rerum natura"* (which is perhaps the first book of physics ever written, about 50 B.C.) felt the need to invoke the goddess of beauty, Venus, to inspire him:

> Mother of Rome, delight of Gods and men,
> Dear Venus that beneath the gliding stars
> Makest to teem the many-voyaged main
> And fruitful lands- for all of living things
> Through thee alone are evermore conceived [3].

However the power of seduction of the siren should not be overstated, both because of its subjectivity and since a Universe too much imbued with symmetry (one of the main canons of beauty) could simply not exist or resemble ours. In fact, as we will see in Chap. 10, a lucky (and yet unexplained) break of symmetry between matter and antimatter allows the former to shape stars, planets and whatever else we observe, while the latter seems to have largely disappeared. So take care: *if you see an antimatter version of yourself running towards you, think twice before embracing* [4], lest you both be annihilated. Another basic break of symmetry in the weak nuclear force will be discussed in Chap. 9, and life, as we know it, could not exist without the asymmetry between D and L-glucose, two stereo isomers of the sugar glucose.

Now that we have gained an idea of what Physics is and of its *ways and means*, let us start in the next chapter our journey, keeping in mind (lest we become disheartened) that, as in the old proverb, attributed to Confucius (551–479 B.C), "the way is the goal".

References

1. M. Tegmark, Our Mathematical Universe: My Quest for the Ultimate Nature of Reality
2. Dante: The Divine Comedy, (Book III, Heaven, Canto 2) translated by Clive James, Picador
3. Titus Lucretius Caro, "De Rerum Natura", Poem, Translated by William Ellery Leonard
4. J. Richard Gott III, Time Travel in Einstein's Universe: The Physical Possibilities of Travel Through Time

Chapter 2
Dramatis Personae (The Actors)

[Ignorance] of the principle of conservation of energy ... does not prevent inventors without background from continually putting forward perpetual motion machines... Also, such persons undoubtedly have their exact counterparts in the fields of art, finance, education, and all other departments of human activity... persons who are unwilling to take the time and to make the effort required to find what the known facts are before they become the champions of unsupported opinions—people who take sides first and look up facts afterward when the tendency to distort the facts to conform to the opinions has become well-nigh irresistible.

Robert Millikan

Abstract Clarity in physics requires precise definitions of terms, such as work and energy, that have fuzzy meanings in everyday usage. Some of the common terms and concepts (e.g. force, velocity, acceleration, momentum) are thus introduced and defined, as an introduction to the material of later chapters, where they will be extensively utilized. Since the book is devoted to readers with only a basic background of mathematics, some of the necessary mathematical tools and concepts are also briefly explained.

2.1 Definitions

R.A. Millikan is undoubtedly right about the many inventors of perpetual motion machines (and incidentally also about their counterparts in several other contexts). It may therefore seem paradoxical that all objects in the Universe are apparently in perpetual motion, whether they are extremely small particles or incredibly large celestial bodies, such as stars or galaxies. To study motion in all its aspects and manifestations, we need to begin with clear definitions. In this Chapter we provide an introduction to some of the more relevant terms and concepts of Mechanics. We will also define a basic tool of the (*Infinitesimal*) *Calculus*,[1] the *derivative*, which is the key to modern mathematics and physics, and explain its meaning (although only

---

[1]Modern calculus is considered to have been developed independently in the 17th century by Isaac Newton and Gottfried Leibniz.

R. Barrett et al., *Physics: The Ultimate Adventure*, Undergraduate Lecture Notes in Physics, DOI 10.1007/978-3-319-31691-8_2

in a very cursory way). Derivatives are also extremely important for physicists, since they are the primary ingredients of *differential equations*, which represent the most customary way of describing the laws of nature in their universality.[2]

A list of definitions is a necessary prerequisite to science, although unfortunately it may be very tedious for the reader. However, without unambiguous definitions, the door is open to sleights of hand (sometimes even involuntary), in which one may advance a thesis by slightly changing the implicit meaning of a term during the course of an argument. A lack of clarity may also lead to distortions of under-standing between participants in a dialogue, due to subjective interpretations of the terminology, so that they may reach different conclusions, and still be logically correct.

Physics is concerned with properties which, at least in principle, are measurable and as such are called *quantities*. For instance, speed, temperature and humidity can be measured, while beauty, usefulness and goodness cannot and therefore are considered *qualities*. This distinction, however, depends on the capability and purpose of the (experimental) scientist. Temperature was not unambiguously measurable before the invention of thermometers, and consequently could not be considered a quantity. Similarly, it is entirely possible, although probably not very useful, to define beauty in terms of measurable criteria, such as the length of the nose or the body mass index (i.e. a person's weight in kilograms divided by the square of height in meters), as has actually been done by some *art theoreticians*, in which case it would become a *quantity* (at least for the people embracing those criteria).

Quantities may have a constant or a variable value: accordingly they are called *constants* or *variables*. Their designation depends on the context in which they are considered: e.g. they might be constant in time but not in space, or vice versa. Let us call x, y and z the space coordinates, corresponding to three arbitrarily chosen, mutually orthogonal directions in the 3D (three-dimensional) physical space. We can introduce time t as an orthogonal fourth dimension, in order to have a more comprehensive 4D space-time. Let us also assume that a given variable f depends on (or is a function of) any n $(1 \leq n \leq 4)$ of the four coordinates of the 4D space-time. Then those n coordinates will define the space of dependence of the variable f.

As an example of such a process, let us consider an orographic map of a given region, i.e. a representation on a plane of the altitude *above sea level* of a land mass by means of contour lines or of colour shades. In this case we have two coordinates (longitude x and latitude y), which yield a 2D space. This is the plane upon which the map is drawn. For each point in the map the elevation e is defined as a function of x and y: i.e., $e = e(x, y)$.

---

[2]In most cases the laws of physics are expressed by means of differential equations, which means that the derivatives of the relevant functions are calculated keeping all variables fixed except the one of interest in the specific term.

More generally the variable f may be a function of n other variables, such as temperature, frequency or density. We may call all of these variables *coordinates*, and define a corresponding (generalized) *nD space*, of which our familiar 3D space and the 4D space-time are but special cases. Just as x, y and z define a *point* in our 3D space, the set of values of the generalized coordinates defines a *point* in the corresponding nD *space*. As an example let us consider the cooking of spaghetti. Here the "coordinates", upon which the quality q of the cooked *pasta* depends, are the cooking time t, the relative quantity of salt s, and the air pressure p: i.e. $q = q(t, s, p)$. A point in the space (t, s, p) *will* define the ideal *parameters* to cook perfect *spaghetti al dente*.

In the following chapters we shall encounter several of the all-important constants of physics. To take just one example here, we may mention the *universal gravitational constant G*, first introduced by Newton [1]. He realized that the motion of planets and the fall of an apple to the ground, as well as the reciprocal attraction of any two massive bodies, all obey the same physical law: namely that the force between them is directly proportional to the product of the masses and inversely proportional to the square of their distance, with a constant of proportionality G. Later we will return to Newton's law, but here we wish to remark that G has a specific value,[3] which is (for what we know) universal. The Universe would be a completely different place if its value were even slightly different. Also, while we normally assume that G has never changed[4] and expect it to remain constant in the future, we have no real proof that this is the case. All we can do here, as in many other cases, is to rely on the so-called *Occam's razor*,[5] which requires us always to choose the *simplest* explanation for the available data, i.e. the most economical one in terms of the assumptions made.

Another basic concept, as we mentioned before, is that of *derivative*, which we define as the instantaneous rate of change of a given function. In other words, if the variable f changes (at a given time) by a quantity Δf in the time interval Δt, its derivative, w, is given by the ratio of these two quantities, provided that Δt is infinitesimal (i.e. arbitrarily small[6]).[7] Using a mathematical notation we write

---

[3]Approximately G = 6.673×10^{-11} N(m/kg)^2.

[4]The possible variability of fundamental physical constants is discussed in Chap. 13.

[5]Occam's razor is an epistemological principle, devised by William of Ockham (c. 1287–1347), which states that, among competing hypotheses that predict equally well, one should always select the one with the fewest assumptions. Other, more complicated solutions may ultimately provide better predictions, but, in the absence of differences in predictive ability, the fewer assumptions, the better.

[6]To be more precise, the derivative is the *limit* of the ratio $\Delta f / \Delta t$, when Δt *tends to zero*.

[7]If the function f depends not only on time (as in the example in the text) but also on other variables, such as the coordinates of the position where the function is evaluated, the derivative may be calculated as a *partial derivative* proceeding as illustrated in the text, but keeping all variables fixed with the exception of one of them. E.g., given a function $f(x, y, z)$ of x, y and z, the partial derivative with respect to y is written $w = \partial f / \partial y$.

$$w = \frac{df}{dt}$$

As an example, consider a traveller driving between Naples and Venice (never mind whether those two cities are the ones in Florida or in Italy). To predict when he/she will arrive, we require the *average velocity*. For a physicist, however, the velocity is defined as the (instantaneous) *displacement* (or shift in *position*) in an *infinitesimal* time interval, i.e. as the derivative of the *position* with respect to time. The velocity defined in this way has a value that may vary continuously, both in magnitude and direction, throughout the entire journey.

Besides the derivative, another ubiquitous tool for physics is the *integral*. Basically an integral is a sum and the symbol used for it, $\int$, is indeed a stretched S, as in Sum. On one side the (indefinite) integral is the inverse of the derivative. Inverting the above formula for the derivative gives:

$$f = \int w dt = \int df$$

On the other side, the sum may be understood by resorting to geometry. To help us, we refer to Fig. 2.1, where the function $w(t)$ is represented.

The product of a given value of w times the interval dt is approximately equal to the area of a strip of width dt. The sum over all wdt's approximately gives the total area under the curve between t_A and t_B. If we let dt become smaller and smaller the number of addends in the sum becomes larger and larger and the result tends to coincide with the actual area. For an infinite number of infinitesimally small addends we obtain the (definite) *integral* of w over t, between t_A and t_B.

Fig. 2.1 The area underneath the curve $w(t)$ between t_A and t_B is the definite integral of the function $w(t)$ between the two values of the independent variable t

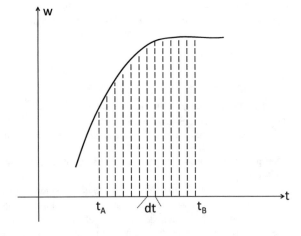

2.2   The Laws of Physics

Likewise, the *acceleration* is defined as the change of velocity in an infinitesimal time interval, i.e. as the (first) derivative of the velocity or the second derivative of the position with respect to time. As we shall see later, an acceleration is always the consequence of an *applied force*; i.e., without an applied force, an object remains forever in its original *state of motion*, whether not moving or moving with constant velocity. This *first law of Mechanics* is counterintuitive, and in fact was clearly stated only about 400 years ago by Galilei [2] as the *principle of inertia*. Since we are continuously surrounded by *fields of force* (mostly gravity), we never experience such an unchanging state of motion.

Even though most students of basic physics may not realize it, the introduction of the concept of force through *the second law of mechanics*, i.e. the famous $F = ma$ (see Chap. 4), arguably one of only two physics formulae known by most non-scientists,[8] requires a non-trivial trick, namely the definition of two new entities (force and mass) by means of a single formula. Of course, once we accept that an acceleration requires a driving force and is proportional to it, the *mass* may be simply defined as the constant of proportionality. However, the mass is also the *source* of gravity. The correspondence of the two interpretations into a single quantity is one of the pillars of Einstein's *Theory of Relativity* (see Chap. 7).

For a complete specification, to define the displacement as the distance an object has moved in a given time interval is not sufficient since the direction in which the movement took place must also be specified. To this end, *vectors* are needed, which are mathematical entities defined by a number (their magnitude in the adopted *units*) and a *direction*. For example, a displacement might be specified as a movement of ten meters in the direction south-to-north. Since our goal in this book is not the application of physics, but rather the illustration of its basic concepts and developments, we will not engage here in the details of vector calculus, which can be found in any basic textbook,[9] but just limit ourselves in the following to providing a few basic examples.

In a *Cartesian*[10] plane, one can define a *position vector r* (by convention, vectors are denoted with bold type) as the *arrow* going from the origin of the coordinate system to the point identifying the current position. Then if one moves from a point P_1 to another P_2, the corresponding displacement *l* is given by the arrow joining the two points P_1 and P_2, and the average velocity v_{av} by the ratio *l*/t, where *t* is the time required for the move. Time is a scalar, and is specified by only one number,

---

[8]The other being Einstein's $E = mc^2$ (see Chap. 7).

[9]E.g. D. Halliday, R. Resnick, J. Walker: Fundamentals of Physics, or Young, Freedman and Lewis Ford: *University Physics with Modern Physics*, or Douglas C. Giancoli: *Physics for Scientists and Engineers with Modern Physics*, as well as many others.

[10]A Cartesian coordinate system is a coordinate system that specifies each point uniquely in a plane by a pair of numerical coordinates, which are the signed distances from the point to two fixed perpendicular axes.

e.g. 2 s or three hours. More precisely, the instantaneous velocity v is given by the derivative of l with respect to t, for which the symbol dl/dt is used. And the acceleration a is defined as the derivative of v with respect to t (i.e. dv/dt).

2.3   The Variables of Mechanics

Translations and *rotations* are defined as displacements along a straight line or around a given circle (i.e. at a constant distance from a given point), respectively. A basic theorem of Mechanics states that every infinitesimal displacement can be decomposed into the *sum* of an infinitesimal translation plus an infinitesimal rotation. As a consequence one can study even the most intricate trajectories, by analysing their two components (translational and rotational) and considering them separately, which obviously simplifies the treatment enormously.

Furthermore, between the two types of movement (translation and rotation) there exists a very useful symmetry. This is our first instance of the abstract *beauty* of physics, which was hinted at in Chap. 1. In order to stress the symmetry we summarize all of the basic variables of Mechanics (the *actors of our game*) in the two columns of Fig. 2.2. Of course students of Engineering in their first course of Physics require a deeper understanding of these variables, and of their applications. For us it is enough here to list the definitions in Fig. 2.2, and provide more details later as required. In the next paragraph, however, we wish to discuss in more detail the rotational equivalent of the basic translational law $F = ma$.

As an example, let us assume that we wish to close a heavy door. Obviously we must apply a force, but where and how to apply it is also important. In fact, let us assume that we pull or push with a force perpendicular to the door itself. If we apply such a force on the side of the door where the hinges lay, we achieve nothing. The easiest way to close the door is clearly to push it on the other side (where the handles are). Why? Because in a rotation the relevant quantity is not the force, but the *torque*, which in the case of a force perpendicular to the door is given, in magnitude, by the product of the force times its *lever arm*. The latter, in this case, is the distance between the point where the force is applied and the side of the door with hinges.

If we substitute the force with the torque and the linear acceleration with the angular one, we obtain the second law of Mechanics for rotational motion, i.e. *the torque is proportional to the angular acceleration.* However, the constant of proportionality will no longer be the mass, but another quantity called the *moment of inertia* (usually denoted with the letter I). For a single point-like body of mass m, I is given by mr^2, where r is the distance of the body from the axis of rotation (i.e. the straight line around which the body rotates). In the case of the door, the axis of rotation is the line of the hinges. The body's *angular velocity* is correspondingly defined as the instantaneous rate of change of its *rotational displacement* (the angle by which it rotates), and its *angular acceleration* as the derivative of its angular velocity with respect to time.

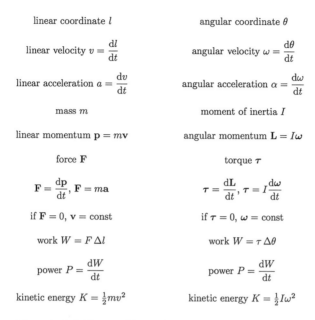

Translational Mechanics   Rotational Mechanics

linear coordinate l angular coordinate θ

linear velocity $v = \dfrac{dl}{dt}$ angular velocity $\omega = \dfrac{d\theta}{dt}$

linear acceleration $a = \dfrac{dv}{dt}$ angular acceleration $\alpha = \dfrac{d\omega}{dt}$

mass m moment of inertia I

linear momentum $\mathbf{p} = m\mathbf{v}$ angular momentum $\mathbf{L} = I\omega$

force $\mathbf{F}$ torque τ

$\mathbf{F} = \dfrac{d\mathbf{p}}{dt}, \ \mathbf{F} = m\mathbf{a}$ $\tau = \dfrac{d\mathbf{L}}{dt}, \ \tau = I\dfrac{d\omega}{dt}$

if $\mathbf{F} = 0$, $\mathbf{v} = $ const   if $\tau = 0$, $\omega = $ const

work $W = F\,\Delta l$ work $W = \tau\,\Delta\theta$

power $P = \dfrac{dW}{dt}$ power $P = \dfrac{dW}{dt}$

kinetic energy $K = \tfrac{1}{2}mv^2$ kinetic energy $K = \tfrac{1}{2}I\omega^2$

Fig. 2.2 Some of the main definitions and laws of Mechanics. The symmetrical correspondence between Translational and Rotational Mechanics may be very helpful, both to understand their meanings and for applications. For the sake of simplicity, we have in some cases limited ourselves to rectilinear or, respectively, circular motion. The generalization to all kinds of trajectories, using vectorial notations, is straightforward. A comprehensive discussion of the laws of Mechanics will be included in Chap. 3

2.4   Conservation Laws

A very important variable of translational Mechanics is the *linear momentum* $\mathbf{p}$, which for an individual body is defined as the product of its mass m times its velocity $\mathbf{v}$. It can be easily seen[11] that, if m is constant, the time derivative of $\mathbf{p}$ is the applied force, i.e. $\mathbf{F} = d\mathbf{p}/dt$. In fact, in Classical Physics, the latter is an alternative formulation to $\mathbf{F} = m\mathbf{a}$. We will see in Chap. 7 that, in the *Theory of Relativity,* the two formulations do not coincide in our customary 3D space because the mass is no longer constant, and only $\mathbf{F} = d\mathbf{p}/dt$ is correct. What happens if no force is applied? In this case $d\mathbf{p}/dt = 0$ and $\mathbf{p}$ is constant. The derivative of a constant vanishes, since by definition a constant does not change, hence the rate of change of the linear momentum is zero.

---

[11]$\mathbf{F} = m\mathbf{a} = m\ dv/dt = $ (if m is constant) $d(mv)/dt = dp/dt$.

What we have just discovered is our first example of a conservation law, the law of conservation of linear momentum. A second example follows immediately, thanks to the above mentioned correspondence between translational and rotational motion. Since the torque, moment of inertia and angular velocity correspond to the force, mass and linear velocity, respectively, the angular momentum is conserved if no torque is present.

Why are conservation laws so important? Because they can immediately explain many *effects* that are easily observed. For instance, if a boat is stationary in water, since no external force is applied its linear momentum is conserved. Then, if a passenger moves forwards, the boat *must* float backwards, so that the total *p* remains zero. Likewise, if spinning dancers wish to increase their angular velocity (i.e. rotate faster), all they have to do is bring their arms and legs as close as possible to their rotational axes to decrease their moment of inertia. As their angular momentum (see Fig. 2.2) must remain constant, since no external torque is present, their angular velocity increases to compensate for the decrease in their moment of inertia.

2.5 Work and Energy

There is a third conservation law, *the law of conservation of energy*, which is extremely important throughout all of Physics. It is a modern version of the famous principle "*Nothing is lost, nothing is created, everything is transformed*" of Antoine Lavoisier,[12] which in another context is known as the *First Law of Thermodynamics* (see Chap. 5). Note the word "*Law*", which implies that it cannot be proved (like most of Physics and in sharp contrast with Mathematics) and might even turn out to be incorrect in as yet unexplored realms of Physics. In fact, its original formulation had to be extended (to include energy arising from matter) when phenomena, such as the annihilation between a particle and its corresponding antiparticle, were discovered.

In daily life the word *energy* can assume different meanings. In classical Physics, energy has an unambiguous definition as the capacity for doing *mechanical work* and overcoming *resistance*. Let us therefore start with the definition of (mechanical) work *W*. If a constant force pulls a body along a straight line for a distance *l*, then the work performed by the force is defined as the product of the two *moduli F* and *l*. More generally, since the directions of the two vectors *F* and *l* may not coincide, the work is given by the product of *F, l* and the cosine of the angle between the directions of *F* and *l*. If the two vectors have the same direction, the angle is zero,

---

[12]Incidentally Lavoisier was guillotined during the French revolution, but not because of his sayings. Among his accusers there was an amateur chemist, Jean Paul Marat, whom Lavoisier had previously rejected as an associate to the Academy of Sciences.

the cosine is 1 and $W = Fl$; if they have opposite directions, the cosine is -1[13] and $W = -Fl$; if they are orthogonal (perpendicular to each other), the cosine is zero and $W = 0$.

A negative value for the work means that the force is resistive and the displacement takes place *in spite* of the force. A vanishing work means that the force is irrelevant to the motion, as when an object is moved across a horizontal floor. In this case gravity, being perpendicular to the floor, performs no mechanical work by itself. It may, however, have other important effects, such as increasing the friction between the object and the floor and consequently increasing our work, if we wish to push the object along.

Of course, such a definition of work does not always coincide with our common perception of the term. In fact, according to this definition, the work performed by a teacher is extremely small, since it is limited only to writing on the chalkboard, plus a little neuronal activity (which involves a negligible amount of work). Even worse, if we ask a porter to carry a heavy suitcase up ten flights of stairs and then we change our mind and ask him to bring it back down to the ground floor, we would owe him nothing for his labour, since the two *works*, up and down the stairs, are equal and opposite (from the viewpoint of the gravitational field) and their sum adds up to zero.

Let us now assume that we accelerate a car up to a certain velocity (say v_f) and then take our foot off the gas pedal. The car keeps moving along the direction of v_f, although its speed decreases due to friction and air resistance. We can explain this effect by saying that the work done by the motor to accelerate the car up to the velocity v_f has been transformed into *kinetic energy* (energy of motion). The kinetic energy has the capacity for doing new work overcoming the friction and air resistance for an additional distance. Similarly, if an object sits on the edge of a table, it is said to possess a *potential energy*, since, by pushing it over the edge, it falls down, enabling gravity to perform work. During the fall the object gains speed and its potential energy is transformed into an equal amount of kinetic energy.

More generally, in all physical processes the *type* of energy (potential, kinetic, heat, electromagnetic, etc.) can be indefinitely transformed, but the total amount of energy does not change. Let us consider another example: a child drops a ball from a given height h, where the ball has a potential energy mgh (m is the mass of the ball and g is the value of the gravitational acceleration on the surface of earth[14]). As the ball falls, the potential energy becomes kinetic energy, which in turn, upon hitting the floor, becomes *elastic energy* with the ball being deformed (slightly flattened). Almost instantaneously the elastic energy becomes kinetic energy again and the ball bounces back up to the original height h. At this point its speed (and kinetic energy) vanishes, but the ball has regained the original potential energy, and the process goes on indefinitely.

---

[13]As the vectors are in opposite directions, the angle between them is $180°$, which has a cosine of -1.

[14]Approximately $g = 9.81$ ms^{-2}.

Or does it? We all know that, even with the most elastic ball and floor, the height *h* at which the ball comes to rest decreases at every bounce, because each time a tiny amount of energy is lost as *heat* (another form of energy). A similar transformation of kinetic energy into heat can be noticed if one strikes an anvil with a hammer: after a few blows the anvil becomes warm to the touch. Again, no energy is lost: the kinetic energy of the hammer disappears, but an equal amount of heat energy is created.

However, in the process something important and *irreversible* occurs. Although the mechanism of transformation between potential and kinetic energy can go on repeatedly, the heat of the anvil can hardly be collected and reused (e.g. to yield again kinetic energy). In other words, heat, being energy, can still produce work but at a lower rate of *efficiency*. In the real world, most (or maybe all) natural processes are irreversible and lead to the gradual transformation of useable energy into less efficient forms of energy. This (so far) never falsified observation is called the *Second Law of Thermodynamics* (see Chap. 5). It can be summarized by paraphrasing George Orwell:[15]

All energies are equal, but some energies are more equal than others.

When it was formulated,[16] the Second Law of Thermodynamics held mostly a practical interest for its application to heat engines. Nowadays, however, due to a later formulation binding it to the concept of an ever increasing *entropy* (see Chap. 5), the Second Law has acquired an almost philosophical dimension. Because of the reinterpretation of the flow of time following Einstein's Theory of Relativity, it has been proposed to use the Second Law for the definition of the arrow of time, as we shall see in Chap. 7.

2.6   Taylor Expansions

To conclude this Chapter, we wish to mention a simple mathematical tool which we will need repeatedly in the following chapters, i.e. *power series expansions*. They can be of great help when dealing with problems for which it is not easy, or even possible, to find solutions in terms of elementary analytical functions. In fact, under some general conditions, which we are not discussing here, they may allow us to find suitable approximations to the exact solution. To explain how they work without using full mathematical rigor, we take advantage of geometric representations.

Let us consider a function *f(x)* in the proximity of a given point *P* and draw the straight line tangent to the corresponding curve in *P*: see Fig. 2.3. The magnified

---

[15]From *Animal Farm* by George Orwell (1947) "All animals are equal, but some animals are more equal than others.".

[16]By Sadi Carnot (1824) and Rudolf Clausius (1850), independently. It can be easily proved that the two formulations are equivalent.

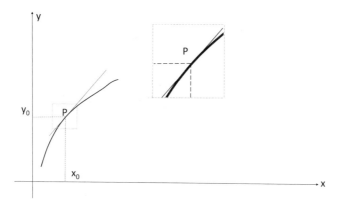

Fig. 2.3 In the vicinity of a point P a curve may be approximated by the tangent in that point

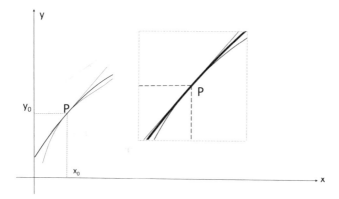

Fig. 2.4 The parabola passing through P and having the same tangent as the original curve is locally a better approximation than the tangent itself

box shows that near to P the difference between the straight line $y(x) = y(x_0) + k(x - x_0)$ where k is a constant, and the curve $f(x)$, is small: the closer to P, the smaller the difference.

An even better approximation may be obtained with a *parabola* passing through P and having there the same tangent as the curve $f(x)$: see Fig. 2.4. The equation of the parabola is also simple since the change in the y value with respect to the tangent is proportional to $(x - x_0)^2$:

$$y(x) = y(x_0) + k(x - x_0) + h(x - x_0)^2,$$

where h is another constant.

We may continue in the same way adding a term proportional to $(x - x_0)^3$, then to $(x - x_0)^4$ and so on. Under the appropriate conditions, each new term will be smaller than the previous ones, yielding a better approximation. It can be proved that an infinite series built in this way is fully equivalent to the function $f(x)$. It is called the *Taylor (power) series*; when the reference point is the origin ($x_0 = 0$) its name is the *Maclaurin series*.

As a very elementary application of these series, let us assume that we wish to calculate the ratio 1/0.99. A simple example of the Taylor expansion is provided by the formula:

$$(1 + x)^n = 1 + nx + n(n-1)x^2/2! + n(n-1)(n-2)x^3/3! + \ldots\ldots$$

where the symbol "!" means *factorial*, e.g. $3! = 3*2*1 = 6$, $4! = 4*3*2*1 = 24$, etc. This expansion is valid for any value, positive or negative, of the exponent n if the absolute value of x is smaller than 1. If we limit ourselves to the first two terms in the series and assume $n = -1$ and $x = -0.01$, we obtain:

$$1/0.99 = (1-0.01)^{-1} = 1 + (-1)*(-0.01) = 1.01$$

Keeping three terms in the expansion, we obtain:

$$1/0.99 = 1.01 + (-1)*(-2)*(-0.01)^2/2 = 1.0101$$

Both 1.01 and 1.0101 represent increasingly better approximations to the value of 1/0.99, whose exact value is 1.01010101 …

The use of the Taylor series to provide successively better approximations to the solution of a problem in physics is a common practice, and in the course of our progress through this book, we will encounter further examples. Quite often the simple first order approximation, where only the first two terms are kept, is accurate enough for the purpose of comparing a theoretical prediction with an experimental measurement. The resultant theoretical estimate may not be 100 % accurate—in fact it may not even be possible to solve the theoretical equations to obtain the precise solution—but then experimental measurements also have errors.[17] If the errors in the Taylor expansion are no worse than the experimental errors, the theory can be meaningfully compared with experiment.

In this chapter we have introduced a few of the important quantities, principles and ideas that underlie physics. Many of them will be revisited in later Chapters. Some terms, as we have seen, have a somewhat different and more specific meaning in physics than in the everyday vernacular. However, physics as a science relies not just on the generation of new ideas and concepts, but also on the rigorous testing of these ideas with controlled and precise measurements and observations. The design and carrying out of such experiments is the task of the experimental physicist. In the next Chapter, we will provide for the lay reader a glimpse into the domain of experimental physics and the methodology of measurement.

---

[17]We discuss measurement errors and their sources in Chap. 3.

References

1. I. Newton, Philosophiæ Naturalis Principia Mathematica ("Mathematical Principles of Natural Philosophy") (1687)
2. G. Galilei, *Dialogue Concerning the Two Chief World Systems* (1631)

Chapter 3
Is Physics an Exact Science?

When you can measure what you are speaking about, and express it in numbers, you know something about it, when you cannot express it in numbers, your knowledge is of a meagre and unsatisfactory kind; it may be the beginning of knowledge, but you have scarcely, in your thoughts advanced to the stage of science.

William Thomson (Lord Kelvin) [1]

Abstract This chapter explains the part played by experimental measurement in physics. Precise measurements necessitate reproducible standards for the fundamental quantities of length, mass and time. These standards have been refined as more accurate measurements of physical quantities have become possible. Physical measurements are always accompanied by an experimental error. Some pitfalls to beware when using experimental data are presented. The different natures of hypotheses, models and theories are explained, and how to distinguish good science from bad.

3.1 Beginnings

In the eyes of the layperson, physics has traditionally been regarded as a more exact science than, for instance, biology which is generally perceived as being descriptive. The avuncular Sir David Attenborough, whose image we have all seen describing and explaining the world of nature on a multitude of TV shows, is arguably the personification of this public perception of a biologist.

However, in recent years, with the advent of gene technology and the use of statistical techniques, biology has moved away from simple observation and description, and come to resemble the physical sciences in the methodology applied to trials and the analysis of experiments. On the other hand, at the current frontiers of physics where objects are very small (fundamental particle physics) or very large (astronomy, cosmology), experiments are difficult and observations hugely expensive. The Higgs boson (see Chap. 10), a crucial component of the theory of matter, was postulated in 1964, but experimental confirmation of its existence had

© Springer International Publishing Switzerland 2016
R. Barrett et al., *Physics: The Ultimate Adventure*, Undergraduate Lecture Notes in Physics, DOI 10.1007/978-3-319-31691-8_3

to wait until 2013 and involved the use of the Large Hadron Collider (LHC) at the European Organisation for Nuclear Research (CERN). The LHC was built in a collaborative project involving thousands of scientists and engineers from around the world at an estimated cost of $10 billion.

The ancient Egyptians, Greeks and Romans were skilled engineers—the Egyptian pyramids, Greek architecture and Roman aqueducts still standing today attest to this fact—but, as we mentioned in Chap. 2, it is generally held that the modern science of physics as we now recognise it began with Galileo in the late sixteenth century. At this time, the teachings of Aristotle and other Ancients were treated with an awe and reverence comparable with that accorded to the Holy Scriptures. Few people were brave enough to question Aristotle's assertion that objects moved only so long as they were pushed, and that as soon as the propelling force was removed the object came to a halt. This, despite the fact that a bow and arrow were hardly unknown, and provide a compelling counter-example.

Aristotle's contention that the velocity of a falling object is directly proportional to its weight was allegedly tested experimentally in 1589 when Galileo dropped two balls of different masses from the top of the Leaning Tower of Pisa and observed their fall. Leaving aside any doubts over the historical authenticity of the story, this experiment contains most of the elements of the modern scientific method. A hypothesis—Aristotle's assertion of the fall velocity being dependent on the object's mass—is tested by direct trial. To carry out the experiment, concepts such as mass and velocity, which we have discussed in Chap. 2, need to be quantified, so that the experimental results can be verified by another experimenter in a different place and time.

It is this methodology that is at the heart of modern physics, and in this chapter we hope to lead the reader to an understanding of the principles involved, with examples of some of its successes and highlighting a few of the pitfalls in the interpretation of experimental results.

3.2   Higher, Faster, Heavier, but by How Much?

Let us imagine that we are standing with Galileo in Pisa at the top of the Leaning Tower to observe the great man dropping balls over the edge. He asserts that one ball is heavier than the other. He produces a set of balance scales and places one ball in each pan. Yes, you agree, one ball is indeed heavier. But by how much? You would like to know the precise weight of each ball so that at a later date the experiment can be repeated. Also, how long will they take to hit the ground and how far will they have fallen? To answer these three questions we need to agree upon standards for length, mass and time against which all measurements can be compared.

Length, mass and time are fundamental quantities in classical physics. Their units are called fundamental units, and in the International System of Units[1] (SI), they are the metre, kilogram and second. (In addition, the SI system contains four more fundamental units which we shall not consider here. These are the candela, ampere, kelvin and mole.) Fundamental units are those from which all other measurable quantities are derived. For instance, we have seen in Chap. 2 that the average velocity is determined by measuring the distance travelled by an object in a specified time. The development of reproducible standards for the fundamental units was an essential prerequisite for the evolution of physics as we know it today. In the next few pages we will touch on a little of this history.

Despite the ancient Greeks having determined the length of the year very precisely in terms of days, at the time of Galileo there existed no suitable device with which small intervals of time could be measured. To tackle this problem Galileo used an inclined plane to slow the fall rate of a rolling ball, and his own pulse and a simple water clock to determine the time for the ball to roll a specific distance. The obvious inaccuracy of these approaches may have been a motivation for his later studies into the motion of pendulums, and their application to the measurement of time. These studies came to fruition in 1656 after his death when Christiaan Huygens, a Dutch mathematician, produced the first working pendulum clock.

Originally the unit of time, the second, was defined as 1/86,400 of the mean solar day, a concept defined by astronomers. However, as earth-bound clocks became more accurate, irregularities in the rotation of the earth and its trajectory around the sun meant that the old definition was not precise enough for the developing clock-making technology. An example of the progress in this technology is the development of the chronometer in the 18th century by John Harrison, which facilitated the accurate determination by a ship of its position when far out to sea, and contributed to an age of long and safer sea travel.

Following further inadequate attempts to refine the astronomical definition of the second, the advent of highly accurate atomic clocks enabled a completely novel approach to the definition of the second in terms of atomic radiation. This form of radiation is emitted when an atom is excited in some manner, and then decays back to its unexcited state. We will learn more of this process in Chap. 9. An example of such radiation is the yellow flare observed when common salt is sprinkled into a gas flame. For some atoms, the frequency of the emitted radiation is very stable and can be used as the basis of time keeping.[2]

The succession from one standard for time to another—from astronomical observations to mechanical oscillations (e.g. the pendulum or balance wheel) to the period of radiation from atomic transitions—occurred because of a lack of confidence

---

[1] SI is the abbreviation from the French: Le Système international d'unités, or International System of Units, and is the modern form of the metric system used widely throughout the world in science and commerce.

[2] In 1967 the second was defined as the duration of 9,192,631,770 periods of the radiation corresponding to the transition between the two hyperfine levels of the ground state of the caesium 133 atom at a temperature of 0 K. This definition still holds.

in the stability or reproducibility of the original standard. But how can we know that it is the original standard and not the new one that is at fault? Why are we so sure that atomic clocks are better at measuring time than a Harrison chronometer? All we can say with certainty is that there are small discrepancies when the two different methods are compared. We come down on the side of the atomic clock because there is more consistency between a plethora of experiments and observations when we use the new standard. This is an application of Occam's Razor (see Chap. 2) which is all very well, provided we are aware of what we have done.

The earliest attempts to standardise a measurement of length are lost in the distant past. Many involved the use of body parts, an advantage in that they were always available when required and seldom lost, but obviously depend on the physique of the experimenter. A cubit was defined as the distance from fingertip to elbow, and the foot and hand were also measures. The latter is still used in expressing the height of horses. The yard was defined as the distance from the tip of King Henry I of England's nose to the end of his thumb. A plethora of other units were also in existence in Britain, including the rod or perch, inch, furlong, fathom, mile, and cable. In time these units became expressed in terms of a standard yard. Various standard yards were used from the time of Henry VII (1485–1509) through to the nineteenth century. In 1824 an Act of the British Parliament decreed a new imperial standard yard which was unfortunately destroyed nine years later by fire. Another new standard was legalised in 1855.

Meanwhile in Europe the French Academy of Sciences was busy defining the metre. Rather than base the measurement on various parts of the human anatomy, they chose to define the metre as one ten millionth of the length of the meridian of longitude passing from the North Pole through Paris to the equator. This length was transcribed to various metal bars over the years, and was not changed, even when the original calculation was found to be in error by 0.2 mm, as a consequence of the neglect of a flattening effect on the earth caused by its rotation.

In 1960, due to the increasing accuracy of length measurements using modern technology, a new definition of the metre, based on a wavelength of Krypton-86 radiation, was adopted by the General Conference on Weights and Measures (CGPM). However, in 1983 this definition was replaced, and the metre is now defined as the length of the path travelled by light in a vacuum during a specified time interval.

Astute readers will realise that this definition now fixes c, the velocity of light in a vacuum, over all space and time at an arbitrary constant. The metre is now defined as the distance travelled by light in 1/299,792,458 of a second, which determines the velocity of light to be 299,792,458 m/s. After several centuries of effort to measure c, future measurements are now rendered superfluous by the stroke of a pen. We will leave to Chap. 13 the implications of this definition on a consideration of the possible variation of the fundamental physical constants, of which c is one, over the life of the universe.

Eventually Britain succumbed to the power of European cultural imperialism, and the Imperial Inch is now defined as 2.54 cm, removing the need (and expense) of maintaining separate standards. However, if anybody believes that the old measures are no longer in use, they might like to participate in a scientific trial on a navy ship, where sailors making observations have been known to record length variously in feet, fathoms, yards, metres, cables, kiloyards, kilometres and nautical miles, and usually don't bother to write down the units. The U.S. received an expensive lesson in the importance of standardising units when the Mars Climate Orbiter space probe was lost in 1999 during orbital insertion due to instructions from ground-based software being transmitted to the orbiter in non-SI units.

The remaining SI fundamental unit that we are considering here is the kilogram. Originally the gram was defined in 1795 as the mass of one cubic centimetre of water at 4C. A platinum bar equal in mass to 1000 cubic centimetres of water was constructed in 1799, and was the prototype kilogram until superseded by a platinum-iridium bar in 1889, which is known as the International Prototype Kilogram (IPK). The IPK is maintained in a climate-controlled vault in Paris by the International Bureau for Weights and Measures (BIPM). Copies were made and distributed to other countries to serve as local standards. These have been compared with the IPK approximately every forty years to establish traceability of international mass measurements back to the IPK. Accurate modern measurements show that the initial 1795 definition of the gram differs by only 25 parts per million from the IPK.[3]

Moves are afoot to redefine the kilogram in terms of a fundamental physical constant and the General Conference on Weights and Measures (CGPM) in 2011 agreed in principle to define the kilogram in terms of Planck's Constant (see Chap. 6). A final decision on the proposed definition is scheduled for the 26th meeting [2] of the CGPM in 2018, so please watch this space.

3.3   Accuracy in Scientific Measurement

Now that we have clarified what we mean when we talk of a metre, kilogram and second, we are in a position to consider the process of scientific measurement.

As Lord Kelvin asserted in 1883, information expressed in numbers always has a greater aura of authority than qualitative descriptions (see citation at the head of this Chapter). A slightly different take on the same topic was expressed by Antoine de Saint-Exupery in *The Little Prince*: *"If you say to the grown-ups: 'I have seen a beautiful house made of pink bricks, with geraniums in the windows and doves on the roof,' they would not be able to imagine that house. It is necessary to say to them: 'I have seen a house worth a hundred thousand francs.' Then they would exclaim: 'My, how pretty it is!'"*

The converse of Kelvin's observation is certainly not true. Just because something is expressed in numbers does not necessarily mean that it is not a lot of hogwash.

---

[3]The Imperial (Avoirdupois) Pound is now defined as 0.45359237 kg.

Every physical measurement has an associated inaccuracy which is a result of limitations in the measuring technique, a lack of control of parameters that influence the final result by an unknown amount, or other factors. An experimental physicist tries to estimate the magnitude of this unknown error, and include that figure in the final result. Following this convention, an experimental result is usually written as x ± y, which means (see Fig. 3.1 and the discussion in the bullet points below) that there is a 68 % chance that the correct value of the measured quantity lies between x − y and x + y. For instance, the velocity of light measured in 1926 by Albert Michelson using a rotating mirror technique [3] was expressed as 299,796 ± 4 km/s. A more recent determination in 1972 by Evenson et al. [4] used laser interferometry and obtained a value of 299,792.4562 ± 0.0011 km/s, clearly a much more accurate result with an estimated error 1/4000th of the 1926 measurement. However, the much more refined, later experiment shows that Michelson's result is still within his quoted error range.

As the interpretation of experimental results is an important part of physics, and leads in large part to its reputation as an exact science, a few words on the treatment of measurement errors might be appropriate here. For instance, to obtain a measurement of the average speed of an object dropped from a tower an experimenter

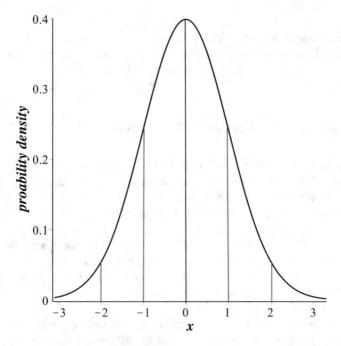

Fig. 3.1 The Normal (or Gaussian) Probability Distribution Function which is followed for the distribution of random errors in experimental measurements. Approximately two-thirds (more precisely, 68 %) of measurements are expected to lie within one standard deviation—which is normally the error quoted with measurements—of the true value (i.e. between −1 and 1 on the graph) and 95 % within two standard deviations (between −2 and 2)

might first measure the height of the tower and then divide this distance by the time taken for the fall. Two different measurements are thus required, one of length and one of time, and both have associated errors which contribute to the error in the final measurement of the average speed of the falling object. It is a waste of resources to use a very accurate process to measure one of these quantities, e.g. a laser to measure the tower height, if the other quantity—time of fall—is not measured with a comparable accuracy. In this case, the error in the time measurement would dominate the final error in the estimate of the average speed.

It is beyond the scope of this book to go into detail on the techniques for estimating experimental measurement error. For further reading, we refer the reader to standard text books on the subject, e.g. Young [5]. Rather, here we wish to make a number of points that should be considered when interpreting the experimental results that one may encounter in scientific journals or popular scientific literature.

Beware of measurements that have no accompanying estimated error. The estimate may have been omitted because the error is embarrassingly large.

Errors in a final measurement (e.g. speed, in the above example) are compounded from the errors in the contributing measurements (length, time) according to the laws of statistics.

Generally the quoted errors are assumed to be random. These random errors may be estimated from knowledge of the apparatus used in the measuring process, or the experiment may be repeated a number of times to determine the statistical distribution of the measurement directly. The repeated measurements obey a bell-shaped (Gaussian) distribution, (see Fig. 3.1), and the quoted error is the Standard Deviation obtained from this distribution.

In addition to random error, there may be a systematic non-random error which has not been detected by the experimenter and which introduces a bias to the final measurement result. For instance, the stopwatch used in the experiment above may have been running slow, and as a consequence the estimated average velocity of the falling object would always be too high.

In deciding whether measurements are consistent with a particular value (e.g. a theoretical prediction) the laws of statistics state that on average two-thirds of the measurements should lie within the quoted error range (one standard deviation) of the prediction and 95 % of the measurements within twice that range.

If many more than two thirds of the measurements do not encompass the prediction in their error range then we can conclude that the experiment does not support the theoretical prediction.

Conversely, if most of the measurements agree and lie within the estimated error range of each other, the agreement may be *too good to be true*—remember that one third of the measurements are expected to lie outside of the quoted range. We should treat such results with caution. The anomaly may be caused by poor estimation of the quoted error, hidden correlations between the measurements so that they are not statistically independent, or some other unknown factor. In any case, proceed with care!

3.4 Measurement of Length in Astronomy

As an example of the difficulties that can arise in a scientific measurement, consider the problem of determining the distance to faraway astronomical objects. On earth the measurement of length is a fairly straightforward process, whether by the use of a tape measure or some more sophisticated tool such as a laser distance measure. However, astronomers are hardly able to run out a tape, and the reflection of laser light and radar waves is only successful for determining the distance to the moon and nearby planets. Nevertheless, it is common to see in the newspapers and popular science magazines that objects have now been discovered at a distance of 13×10^9 light-years.[4] This is an enormously large distance. How are such measurements possible?

A history of the measurement of astronomical distances could easily fill a monograph on its own, and we have no intention of attempting such a task here. Nevertheless, a brief summary of some of the underlying principles is illustrative of the way physicists (or in this case, astronomers) proceed when faced with an apparently intractable problem.

Nearby stars observed from earth appear to move against the background of distant stars as the earth circles the sun. This is an example of parallax, the effect that gives rise to stereoscopic vision in humans because of the slightly different pictures received by our forward-facing separated eyes. A star with one arc-second of observed parallax, measured when the earth is on opposite sides of the sun, is said to be at a distance of 1 parsec. The parsec is the standard unit of distance in astronomy. The distance to the star in parsecs is the reciprocal of the measured parallax in arc-seconds. To convert from parsecs to more conventional units we need to know the distance of the earth from the sun. This distance is defined as the Astronomical Unit (au) and must be measured independently. Again nothing is simple, as estimates of the au are complicated by effects such as relativity due to the motion of the observers.[5]

In the early '90s, the Hipparcos satellite was used to take parallax measurements of nearby stars to an accuracy much greater than possible with earth-bound telescopes, thereby extending the distance measurements of these stars out to ~ 100 parsecs (~ 300 light-years). How can this range be further extended?

The first step involves the comparison of a star's known luminosity with its observed brightness. The latter is the brightness (or apparent magnitude) observed at the telescope. It is less than the intrinsic luminosity (or absolute magnitude) of the star because of the inverse-square attenuation with distance of the observed

---

[4]1 light-year = 9.4607×10^{15} m, i.e. 9461 Billion km.

[5]The Astronomical Unit (au) is currently defined as 149,597,870,700 m, which gives 1 parsec = 3.26 light-years.

radiative energy.[6] If we know how bright the star is intrinsically, we can estimate its distance away using the inverse square law.

From measurements on stars in our galaxy within the 300 light-year range it was discovered that stars of similar particular types have the same absolute magnitude. If stars of these types are observed outside the range where parallax measurements are observable, we can estimate their distance by assuming their absolute magnitude is the same as closer stars of the same type, observe their apparent magnitude, and use the inverse square law to compute their distances. The distance to another galaxy can be inferred from the distance to particular stars within it.

A third approach for measuring the distance to faraway objects came from the observation of Edwin Hubble that the light spectra observed from distant galaxies were displaced towards the red end of the spectrum. This phenomenon is analogous to the Doppler Effect, which produces a drop in pitch of the sound from a train as it passes and recedes from an observer. Hubble concluded that the distant galaxies were moving away from us, and that the fainter, more distant galaxies were moving faster than those closer to us. This is the characteristic of an expanding universe. The red shift is proportional to the distance to the galaxy, and the constant of proportionality is known as Hubble's constant.[7] From Hubble's constant and the observed red shift we can calculate the distance to the farthest astronomical objects. So how do we estimate Hubble's constant?

Just as there is a region of overlap between where parallax measurements and luminosity measurements are possible, there is another region of overlap between luminosity and red shift measurements. A comparison of the two sets of observations enables an estimate of Hubble's constant. It sounds simple, but decades of work have been undertaken to refine the accepted value of Hubble's constant. These estimates have fluctuated quite considerably. The estimated age of the universe is directly related to the Hubble constant.

The uncertainty that is involved in astronomical estimates was highlighted by the eccentric 20th century mathematician, Paul Erdös, who when asked his age, declared he must be about 2.5 billion years old because in his youth the earth was known to be 2 billion years old and now it is known to be 4.5 billion years old.

The problem of estimating very small (i.e. sub-atomic) distances is another confronting problem in measurement that we will not discuss further here.

---

[6]When an object radiates uniformly in space, one can envisage the energy being carried out on ever expanding spherical wavefronts. As the energy is distributed uniformly over the spherical wavefront, its density is reduced as the area of the wavefront increases. The wavefront area is proportional to the square of the sphere's radius, hence the energy intensity of the radiation falls away as the inverse square of the distance to the radiator.

[7]Note the implicit assumption here that the expansion is uniform.

3.5   The Path to Understanding

Now that we have some idea of what is involved in the experimental and observational processes, and have learned to treat experimental results with some cautious respect, it is appropriate to examine the aims of an exact science. The collection of experimental data is an important component of scientific enquiry, but the ultimate goal is an understanding of the physical processes underlying the observations. Such an understanding leads not only to an explanation of the observed results, but also to a quantitative prediction of the results of experiments not yet undertaken. This predictive ability is the distinguishing characteristic of good science.

The first step in the understanding of a physical process, according to what is generally known as "the scientific method", is usually the establishment of a testable hypothesis. By "testable" we mean that the hypothesis leads to predictable outcomes that can be subjected to an experimental test. For instance, Galileo tested the hypothesis due to Aristotle that a body's rate of fall depends on its mass. His work disproved the hypothesis and led to the birth of Newtonian mechanics. If there is no way a hypothesis can be tested experimentally, even if that test may lay some way in the future and be of an indirect nature, it has little scientific value. For instance, the concept of atoms can be traced back to the ancient Greeks, and formed the basis of modern chemistry even before individual atoms could be directly observed in scattering experiments.

Wolfgang Pauli, one of the greats of 20th Century Modern Physics, disparaged untestable theories as "not even wrong". This, in his eyes, was a far worse characteristic than being wrong, for the experimental testing of wrong theories often leads to unexpected new breakthroughs. The Steady State Theory of the Universe (see later), although now believed wrong, inspired many experimental and theoretical investigations. Pauli must have experienced a crisis of conscience when he predicted in 1930 the existence of the neutrino, a particle with no (or very little) mass and no charge (see Chap. 10). "*I have done a terrible thing,*" he wrote. "*I have postulated a particle that cannot be detected.*" History has proved him wrong; several variants of the neutrino have since been discovered, as we will see in Chap. 10.

Much of science is a deductive process, making use of rigorous mathematical logic. A myth of popular psychology has it that these processes occur in the left hemisphere of the brain, whereas the "creative" intuitive processes that are the basis of art occur in the right cerebral hemisphere [6]. Such an assertion shows a lack of understanding of the scientific method. The formation of a hypothesis is not deductive, but intuitive. Most scientists have their Eureka moments, when a new idea or concept suddenly pops into their heads while they are walking the dog, washing the dishes or languishing, like Archimedes, in their bath. The deductive component comes in deducing the consequences that should follow from the hypothesis.

As hypotheses are postulated and tested experimentally a growing understanding of the physical process under investigation develops. This knowledge can be further crystallised into a "model", or a scientific "theory".

The term "model" brings into mind a physical structure, such as a model aircraft. However, scientific models are usually mathematical. As we shall see in Chap. 9, the Bohr-Rutherford model of the atom envisaged the atom as a miniature planetary system with a heavy, positively-charged nucleus at its centre and the negatively-charged electrons orbiting about the nucleus. With a few assumptions relating to the stability of the orbits, this model was highly successful in explaining the spectra of light radiated by the simpler atoms when they become excited. However, the physical structure of an atom is now known to be quite different from the Bohr-Rutherford model.

Models form a valuable function in modern physics, and examples will be given in later chapters of models applied to various physical processes. However, the reader would do well to apply caution when considering the results of modelling. Useful models, such as that of Bohr, require few input assumptions and predict with considerable accuracy the outcome of a variety of precise experiments. Poor models have many parameters that must be tuned carefully to account for past experimental data. Their predictions, when tested with new observations, are often in error until the parameter set is enlarged and re-tuned. Such models may be dubbed "Nostradamus models", as in the same manner as the writings of Nostradamus, they are only successful in "predicting" what has already taken place.

When our understanding of a physical process has reached a deeper level than can be obtained with modelling, it is usually formulated in terms of a "theory", e.g. Newtonian mechanics, or the theory of electromagnetism based on the work of Maxwell and others (see Chaps. 4 and 6). A theory is not something whimsical, as the common English usage of the word would imply, but a framework built with mathematical logic from a small number of physical "laws". In the same way that geometry is constructed by deductions from a small number of axioms postulated by Euclid, so is the science of classical mechanics, which so accurately explains the dynamics of moving bodies in the everyday world, based on deductions from laws of motion postulated by Newton and others.

Every test of a prediction of a physical theory is a test of the underlying physical laws. Some, such as the law of conservation of angular momentum (see Chap. 2), have been found to have a validity extending from the sub-microscopic world of atoms to the farthest galaxies. In some cases, even laws that have stood the test of experiment for centuries, may need modification when they are applied to regions of physics that were not envisaged at the time of their formulation. For instance, the mechanics of Newton gives way to the relativity theory of Einstein for objects travelling at speeds near to that of light. However, Einstein's relativity yields the same predictions as Newton's for the velocities that are encountered in everyday life. Newtonian mechanics can be considered to be a very accurate approximation of Einsteinian relativity for everyday objects, and is still used in preference to Einstein's theory for these because it is mathematically much simpler to apply.

Occasionally two apparently different theories appear to describe experimental observations with equal accuracy. Such was the case with the Matrix Mechanics of Heisenberg and the Wave Mechanics of Schrödinger, both of which accurately predicted observations in the atomic domain. In this case it was discovered that the two theories were in fact equivalent, with the laws of one capable of being derived from the other theory, and vice versa. Today the two approaches are combined under the name of Quantum Mechanics (see Chap. 8).

It is with the rigorous application of well-established theories that the predictive power that has given physics its reputation as an exact science comes to the fore. Quantum Electrodynamics was developed in the 1920s and is a theory describing the interaction of electrically charged particles with photons, e.g. when an electron emits radiation when decaying from an excited state in an atom. Predictions made with this theory have been verified to an accuracy of ten parts in a billion. This is equivalent to measuring the distance from London to Moscow to an accuracy of 3 cm, which is precision indeed.

3.6   Caveat Emptor!

Now that we have an idea of the scientific method and the aims of physics, it is probably appropriate to spend a page or two on the human side of the discipline. Physics is carried out by normal men and women who are subject to the same character traits as the rest of the population. These include ambition, egotism, obstinacy, greed, etc. In some cases, this human element can have an impact on the way that the science is pursued.

For instance, if one has invested a great deal of one's time and energy into the development of a scientific model or theory, it is understandable if one does not greet evidence of its overthrow with alacrity. This attitude is hardly new. Pythagoras held the belief that all phenomena could be expressed in rational numbers (i.e. integers and fractions). A widely circulated legend, probably an academic urban myth, is that Pythagoras drowned one of his students when the unfortunate fellow had the temerity to prove that the square root of two was not expressible in rational numbers.

If doubt exists about the authenticity of the Pythagorean legend, the animosity between Newton and a contemporary, Robert Hooke, is well established. Newton is alleged to have held off the publication of his book on Optics until after Hooke's death so that he could not be accused by Hooke of stealing his work. Hooke is remembered for little today apart from his studies on elasticity. However, he was perhaps the greatest experimental scientist of the seventeenth century, with work ranging over diverse fields (physics, astronomy, chemistry, biology, and geology). Newton has been accused of "borrowing" Hooke's work, and destroying the only portrait of him that existed [7].

Another more recent feud occurred between the engineering genius, Thomas Edison, and Nikola Tesla [8], the inventor of wireless telegraphy and the alternating current. The latter had the potential for, and was ultimately successful in, replacing the use of direct current for home power supplies. As Edison had many patents on the application of direct current, Tesla's work threatened his income. The source of his rancour is thus easy to see.

In the 1960s two competing theories existed side by side to explain the origin of the universe (see Chap. 11). These were the well-known Big Bang Theory and the Steady State Theory, which was propounded by Sir Fred Hoyle, Thomas Gold and Hermann Bondi [9]. The basic premise of the latter was that the universe had no beginning, but had always existed. Matter was being continuously created in intergalactic space to replace that dispersed by the observed expansion of the universe. Presentations by the adherents of the rival theories made scientific conferences at the time entertaining, and sometimes heated.

Eventually the disagreement was resolved using the approach pioneered by Galileo, i.e. observation and measurement. The *coup de grace* for the Steady State Theory occurred with the discovery in 1965 of a background microwave radiation [10] pervading the universe, which had exactly the temperature predicted by the Big Bang Theory. Despite the growing evidence against the Steady State Theory, Fred Hoyle carried his belief in its veracity to the grave.

The purpose of the last few paragraphs is not to disparage physicists of the past, but simply to draw attention to the fact that scientists are subject to the same human frailties as everyone else, and this can impact on their scientific objectivity.

Very strong evidence indeed is required to overturn long-held views, models and theories. This is as it should be, but sometimes errors creep in. For instance, up until the 1950s it was widely held that the laws of physics do not distinguish between left and right. In other words, it is not possible to distinguish the world as viewed through a looking-glass from the real one, Lewis Carroll notwithstanding. When it was proposed by Tsung Dao Lee and Chen Ning Yang that an asymmetry between left and right existed for a particular type of nuclear force known as the weak nuclear interaction, the experiment to verify their hypothesis was performed by Madame Chien-Shiung Wu within a few months.

Why had no one performed such an experiment earlier? Well, in fact they had. Richard Cox and his collaborators had carried out such experiments [11] in 1928, nearly three decades before Madame Wu, but they had attracted little attention. The significance of their work was not understood, even by the authors, so ingrained was the belief by physicists in the left-right symmetry of physical laws. It is a very human trait for scientists to self-censor their experiments, and dismiss as an aberration any experiments that produce results that stray too far from established beliefs.

So how should a non-scientist approach the technical journals and popular scientific literature? With respect, and caution. As we will see in following chapters, there is a vast quantity of innovative and brilliant research work out there, but there is also a lot of junk science, which can be as dangerous to one's well-being as junk food. We hope that this chapter has given the reader a few hints for discriminating between the two.

References

1. W. Thomson (Lord Kelvin), Lecture on "Electrical Units of Measurement" (3 May 1883)
2. Website: http://www.bipm.org/utils/common/pdf/CGPM-2014/25th-CGPM-Resolutions.pdf
3. A.A. Michelson, Measurement of the velocity of light between Mount Wilson and Mount San Antonio. Astrophys. J. **65**, 1–22 (1927)
4. K.M. Evenson, J.S. Wells, F.R. Petersen, B.L. Danielson, G.W. Day, R.L. Barger, J.L. Hall, Speed of light from direct frequency and wavelength measurements of the methane-stabilized laser. Phys. Rev. Lett. **29**, 1346 (1972)
5. H.D. Young, *Statistical Treatment of Experimental Data* (McGraw-Hill, 1962)
6. K. Cherry, Left Brain vs Right Brain: Understanding the Myth of Left Brain and Right Brain Dominance (2015), http://psychology.about.com/od/cognitivepsychology/a/left-brain-right-brain.htm. Accessed 21 May 2015
7. Website: http://www.iop.org/news/12/jan/page_53418.html. Accessed 21 May 2015
8. S. Patrick, Nikola Tesla: Imagination and the Man That Invented the 20th Century, Kindle ebooks (2013)
9. F. Hoyle, A new model for the expanding universe. Mon. Not. R. Astron. Soc. **108**, 372 (1948)
10. A.A. Penzias, R.W. Wilson, A measurement of excess antenna temperature at 4080 Mc/s. Astrophys. J. Lett. **142**, 419–421 (1965)
11. R.T. Cox, C.G. McIlwraith, B. Kurrelmeyer, Apparent evidence of polarization in a beam of ß-rays. Proc. Natl. Acad. Sci. U.S.A. **14**, 544–549 (1928)

Chapter 4
Newton and Beyond

If an elderly but distinguished scientist says that something is possible, he is almost certainly right; but if he says that it is impossible, he is very probably wrong.

Clarke [1]

Abstract The development of mechanics from Galileo to Newton is presented, leading up to Newton's theory of gravity. Geometrical and physical optics are introduced and explained. The formulations of Analytical Mechanics due to Lagrange and Hamilton are presented as examples of how different theoretical approaches can be applied to the same problem, and as a precursor to Quantum Mechanics, which is discussed in a later chapter.

4.1 It's All Been Done!

From Chaps. 2 and 3 we have obtained a basic understanding of the methodology of physics, particularly in relation to the precision of its terminology, the nature of physical theories and the process of observation and experiment. Now the time has come for us to see this methodology in action.

At the end of the 19th Century, physics had reached a stage in its development when all of the outstanding physical problems appeared to have been satisfactorily resolved. The fields of mechanics, optics, and thermodynamics, were surely all done and dusted, and the jewel in the crown of classical physics, the unification by Maxwell of electricity and magnetism into the theory of electromagnetism, had rounded off the 19th Century for physicists in a glow of self-approbation.

So enthused was Lord Kelvin that he proclaimed to the British Association for the Advancement of Science in 1900 that *"there is nothing new to be discovered in physics now. All that remains is more and more precise measurement."* Or so it is alleged. Certainly, the view expressed by Kelvin appears to have been wide-spread at the time.

© Springer International Publishing Switzerland 2016

R. Barrett et al., *Physics: The Ultimate Adventure*, Undergraduate Lecture Notes in Physics, DOI 10.1007/978-3-319-31691-8_4

However, the gift of augury is not necessarily accorded even to the great, and as Science Fiction writer Arthur C. Clarke aptly remarked in the citation that opens this chapter, many a would-be scientific prophet has lived to regret rash statements.

Shortly after Kelvin's unfortunate prophecy, the launch of the 20th Century heralded the development of the theories of Relativity and Quantum Mechanics, both of which turned classical physics on its head and led to what is now termed "*Modern Physics.*" However these ideas did not emerge, fully formed, like the goddess Aphrodite rising from the sea. Rather, they developed from the solid foundations laid by the scientific greats of the past, and it is the purpose of the next three chapters to explore these past achievements. In later chapters we will discover how drastic modifications were forced on these nineteenth century theories by developments in the so-called Golden Age of Physics, which took place in the first half of the twentieth century.

4.2 Newton Stands on the Shoulders of Giants

"*If I have seen further, it is by standing on the shoulders of giants*". So wrote Isaac Newton in a letter [2] to his arch-rival, Robert Hooke in 1676. He was most likely paraphrasing Bernard of Chartres, a 12th century scholar, who compared us (the moderns) with dwarves riding on the shoulders of giants (the Ancients), and suggested that it is for this reason alone that we are able to see further than our antecedents. The expression served also as a sarcastic rejection by Newton of the notion that he had exploited Hooke's work, for Hooke was a sickly man, and certainly not a giant. To set the scene for what is to come in this chapter, a quick review of the work of these "giants" is appropriate at this point.

In 1638, Galileo published his final book, *Discourses and Mathematical Demonstrations Relating to Two New Sciences*. In it he describes the motion of falling bodies, projectiles, the concept of the relative motion of two bodies, and some fundamental ideas about the pitch of vibrating strings, friction and infinity. Following after Galileo, the end of the 17th Century saw a blossoming of studies into the motion of objects, both earthbound and celestial. Christiaan Huygens, John Wallis and Gottfried Leibniz introduced concepts such as the *Conservation of Momentum* and the *Conservation of Energy*, which were discussed in Chap. 2.

Earlier in the century, between 1609 and 1619, based on the meticulous observations of Tycho Brahe in Prague, Johannes Kepler deduced his three laws of planetary motion.[1] Kepler's laws are specific and apply to a very narrow field, but they marked a turning point in the application of science to a celestial problem, and a victory for the heliocentric model of the solar system propounded by Copernicus.

These laws may be stated as follows:

---

[1]A recent historical novel, *Kepler*, by Mann-Booker prize winning novelist, John Banville, describes the life of Kepler and sets the atmosphere of the period in which his work took place.

Law 1   the orbit of every planet is an ellipse with the Sun at one of the two foci;
Law 2   a line joining a planet and the Sun sweeps out equal areas during equal
        intervals of time;
Law 3   the square of the orbital period of a planet is proportional to the cube of the
        semi-major axis of its orbit.[2]

Following on from the discoveries of his predecessors, Newton began a series of investigations that culminated in 1687 with the publication of his masterpiece: *Philosophiae Naturalis Principia Mathematica* (Mathematical Principles of Natural Philosophy). In this book, a pinnacle of rational scientific thought, Newton introduced his three laws of motion and the principle of gravitation, which explain celestial motion. They are still today the starting point for most courses in physics, and essential knowledge for anybody desirous of even a cursory understanding of the physical world about them. Although we have discussed some of these principles in Chap. 2, for clarity in what follows we reiterate them here.

The first law is usually written: *every body continues in its state of rest, or of uniform motion in a straight line, unless it is acted upon by an external force.* With this law, Newton is clarifying the work of Galileo and laying to rest forever Aristotle's idea that objects come to rest when the driving force is removed from them.

The second law gives us a quantitative description of what happens when an external force is indeed applied to any object: the object is then accelerated along the line of the applied force and the magnitude of the acceleration is inversely proportional to the mass of the object. As we have seen in Chap. 2, this law is expressed as $\mathbf{F} = m\mathbf{a}$, where $\mathbf{F}$ is the applied force, m is the mass of the object and $\mathbf{a}$ is the acceleration of the object. $\mathbf{F}$ and $\mathbf{a}$ are vectors. The first law is actually a special case of the second law. If we choose $\mathbf{F} = 0$ in the second law we see that the acceleration of the object is zero, and the object's state of motion is therefore unchanged. If at rest, it stays at rest; if in motion it continues along with the same velocity.

The first and second laws accord with our everyday experience in that it is hard to set heavy objects moving (inertia), but once they are moving it is hard to stop them (momentum). Aristotle's principle might be expressed as $\mathbf{F} = m\mathbf{v}$, which would indicate that if we remove the external force, the object stops immediately. Newton realised that the bringing to rest of a freely moving object was the result of the external forces of friction and air resistance, not the removal of the driving force. If we could place a moving object into an environment where friction and air resistance are removed, the object would continue its motion forever. The recent

---

[2]The orbital period is the time taken for a planet to complete one orbit of the sun; the major axis of an ellipse is the longest diameter which is a straight line running from a point on the ellipse through the centre and both foci to a point on the opposite side of the ellipse; a semi-major axis is half of the major axis.

exit of the space probe Voyager 1 from the solar system thirty-five years after its launch from Earth is a modern demonstration of the truth of Newton's ideas.

Newton's third law states *that for every action, there is an equal and opposite reaction.* It is somewhat more subtle than the first two laws. In one sense it is a clarification of the requirement in Law 1 that the driving force is external to the body being propelled. You can stand in a sailboat and push as hard as you like on the mast but you will not propel the boat one millimetre forwards because your feet exert a rearward force on the bottom of the boat equal and opposite to the force exerted by your push on the mast. You would be better off turning around and blowing over the stern, in which case there would be a "reaction", if ever so slight, to the "action" of the force driving your breath. This is the principle behind the rocket ship and the jet engine.[3]

The third law also explains the "kick" that occurs when a firearm is discharged. In Chap. 2 we introduced the concept of conservation laws, which play an important role in physics. It can be readily shown that the Laws of Conservation of Momentum and Angular Momentum are a consequence of Newton's third law, and the Law of Conservation of Energy arises from Newton's second law.

However, it turns out, as we will see in later chapters, that these conservation laws are more fundamental than Newton's laws and apply in domains such as Atomic Physics where Newton's laws are inapplicable.

Even in the everyday world there are instances where Newton's laws are inapplicable. An experimenter sitting on a carousel would notice that there are effects due to the rotation which complicate the application of Newton's laws, and forces such as centrifugal force and the Coriolis force arise. When we are standing outside the rotating carousel, we may dismiss these forces as "fictitious" or "pseudoforces" due to the rotary motion. However, we spend our life on a rotating body (the earth) and the effects of centrifugal force (e.g. the equatorial bulging and flattening of the earth at the poles) and the Coriolis force (hurricanes) are anything but fictitious. It is customary to add the caveat that Newton's laws only apply in an inertial frame of reference, but if a physicist is asked to define an inertial frame of reference, the usual reply is that it is a frame of reference in which Newton's laws apply. One can think of a non-inertial frame of reference as being one that is undergoing some form of acceleration.

Another attribute of Newton's laws is that they are time-reversible. If one were to view a movie running backwards in time, the same laws of physics would apply as in the real world. It might look strange to see an object rising from the floor and floating up onto a table. However, if you were to work through the equations of motion with the final velocity of the object just before impact reversed, you would

---

[3]It is a popular misconception that the air blown out by a jet engine meets a resistance from the atmospheric air behind the engine, and that as a consequence the engine is thrust forward in the same manner that skaters pushing on a wall can thrust themselves away from it. This is incorrect. A rocket ship works perfectly well in deep space where there is no atmosphere. The forward thrust is a consequence of Newton's third law.

describe the events in the backwards-running film accurately. We will see later that time-reversibility is a requirement of physical laws for individual bodies.

4.3   Newton's Law of Gravity

After expressing his three laws of motion, Newton proceeded in *Principia* to lay out his theory of gravitation. Everybody is aware of the apocryphal story that the idea of gravity came to Newton after he observed an apple fall from a tree. However, Robert Hooke had a different opinion of the law's origin and when Newton propounded his theory to the Royal Society in 1686, Hooke immediately claimed that Newton had plagiarised the law from him.

Putting aside a controversy which still continues to this day, the law of gravity can be stated mathematically (as we saw in Chap. 2) with the force **F** being such that its modulus F is given by:

$$F = G\frac{m_1 m_2}{r^2}$$

where m_1 and m_2 are the two masses, and r is their separation. The force **F** is exerted on each mass, and its direction is towards the other mass,[4] i.e. it is an attractive force, tending to move the two masses towards each other. G is a constant that is known as the Universal Gravitational Constant (see Chap. 2) and is one of the most important constants of the universe. G will be discussed further in Chaps. 7 and 13.

The law of gravity, combined with the three laws of motion, explained most of the astronomy known at the time. Kepler's laws can be deduced from Newton's.[5] The orbits of planets and their masses could be, and were, calculated. A web of gravitational pulls between stars, planets, comets and other celestial objects holds the universe together, and it is the same force that holds our feet on the ground.

The incredible scope of Newton's laws is apparent, and is in stark contrast with the limited nature of Kepler's three laws. Mechanics, as we know it today, came into existence with the publication of the *Principia*.

Although the importance of Newton's *Principia* was quickly recognised, the book was not without its critics. In England it was argued that Newton had removed any place for God in the universe, and was promoting atheism. (It should be remembered that Newton himself was devoutly religious and in later life gave up

---

[4]This is an example of the third law.

[5]As we discussed in Chap. 2, the aim of physics is to explain nature with a minimum of "laws", in the same way that mathematics is constructed from a minimum of axioms. Kepler's laws can be derived from Newton's and are therefore not considered fundamental. In practice, they may still be used if they are easier to apply in a particular example.

physics for theology). In France and Germany, Descartes and Leibniz objected that Newton hadn't explained the nature of gravity, which appeared as almost super-natural in his work.

In a second edition of the *Principia* published in 1713, more than a quarter of a century after the original, Newton addressed some of these criticisms. He admitted that he did not understand the nature of gravity, or how it exerted a pull on objects across empty space. His reply was similar to that of Samuel Beckett who, when asked to explain who Godot was in his absurdist classic, *"Waiting for Godot"*[6] is alleged to have replied: *"if I knew the answer to that question I would have put it in the play."* In a new addendum, Newton rejected the idea that the *Principia* pro-moted atheism, maintaining that the order that his laws found in the universe was a sign of the presence of a creator, who was beyond human understanding.

The three hundred years that have elapsed since the *Principia* first appeared have bred a familiarity with the concept of gravity that has dulled our amazement at the idea that a force can act at a distance through empty space. We now take gravity for granted. As we will see in Chap. 7, Einstein proposed a different view of gravity which sees its effects arising through the distortion of space-time by the presence of a mass. Whether this concept is easier to grasp than the idea of action at a distance, which so troubled Descartes and Leibniz, is a matter for the beholder.

In the *Principia*, Newton proposes his laws and deduces their consequences using precise mathematical logic. He finds that these consequences agree with the results of observations, in some cases providing an explanation for hitherto unex-plained effects (e.g. tides and the orbits of comets). He does not speculate on the nature of gravity. In adopting this approach he has taken the baton from Galileo in laying down the methodology of the science of physics. Physics answers questions beginning with how, when and where. Questions beginning with why are best left to philosophers.

In the following chapters we will be introduced to many concepts which are counter-intuitive and puzzling, and yet meet the test of experimental verification, and the challenge of Occam's Razor. One wonders whether in three hundred years' time, humankind will be as blasé to these 20th and 21st Century ideas as it is today to gravity.

4.4 Let There Be Light

Another branch of physics in which Newton made a ground-breaking contribution is the science of optics (or the study of the properties of light). The importance of light to human vision has ensured that its nature has remained a subject of spec-ulation and investigation throughout recorded history. The earliest Egyptian and

---

[6]*"Waiting for Godot"* is a play about two tramps waiting for a third character, Godot, who never shows up.

Mesopotamian lenses date from 700 BC, and the ancient Greeks had two competing theories of vision.

The first of these, an intromission theory advocated by, among others, Aristotle maintained that objects cast off copies of themselves that are captured and interpreted by the eye. The second, emission theory, supported by Plato and Euclid, held that vision was the result of rays emitted by the eyes.

The Greek ideas were further developed in the Muslim world during the Middle Ages. A 10th Century Persian mathematician, Ibn Sahl, described the process of refraction and formulated a law equivalent to what later became known as Snell's Law (see below). Meanwhile, the practical side of optics, lens making, proceeded on an empirical basis without recourse to much in the way of a reliable physical theory. The first wearable eyeglasses were produced in Italy in the thirteenth century. Spectacle-making centres developed in Germany and the Netherlands, and work there led to the development of the compound microscope in 1595 and the refracting telescope in 1605. Kepler's interest in astronomy and the telescope sparked him into a study of the underlying optics of these instruments. He studied lenses, curved mirrors and the pinhole camera, and used his knowledge to design an improved telescope.

Kepler's work was further developed in the mid-17th century by René Descartes, who replaced the Greek theories with the modern idea that light is emitted by sources. Newton put forward a corpuscular theory of light, which asserted that a ray of light is composed of discrete particles. His work led to what is now known as "*geometrical optics*". Newton's theories were ridiculed by his *bête noire*, Robert Hooke (who else?), and Christiaan Huygens, who both advocated a wave theory of light. Following Hooke's death, Newton's fame and his publication of *Opticks* in 1704 ensured that the corpuscular theory of light gained general acceptance.

4.5 Geometrical Optics: The Corpuscular Theory of Light

Geometrical optics is the study of the reflection of light from mirrors—curved and flat—and its refraction at the interface between two transparent media, e.g. air and glass. The common spectacle lens is formed by two different refracting surfaces, the air-glass interface and the glass-air interface of the front and back surfaces of the lens respectively. However, the lenses of complex instruments such as modern cameras have many refracting interfaces. An understanding of geometrical optics is essential for the design of these instruments.

The foundations of geometrical optics are two important laws: the laws of reflection and refraction. The first states that the angle of incidence of a ray of light on a reflecting surface[7] is equal to the angle of reflection.

---

[7]We are dealing here with "specular" reflection which occurs at a shiny surface, such as a mirror, and not the diffuse reflection from a surface such as frosted glass.

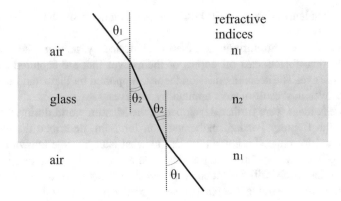

Fig. 4.1 Refraction of a ray of light through a block of glass, according to Snell's law. The angles θ_1 and θ_2 are the angles of incidence and refraction at the air-glass interface, and vice versa at the glass-air interface

These angles are measured with respect to a perpendicular drawn from the surface at the point of contact of the ray with the reflecting surface. In essence the law states that a ray hitting a surface orthogonally (i.e. at a zero incidence angle) is reflected back along its path, while one hitting at a skimming angle is reflected near to the surface of the mirror.

The law of refraction, also known as Snell's law after Willebrord Snell, a 16th Century Dutch astronomer and mathematician, states that the angles of incidence and refraction (see Fig. 4.1) are related by the formula:

$$n_1 \sin \theta_1 = n_2 \sin \theta_2$$

where θ_1 is the angle of incidence and θ_2 is the angle of refraction. n_1 and n_2 are parameters inversely proportional to the speed of light in the two media (air and glass) and are known as their refractive indices. The refractive index of air is usually defined to be unity.[8] Newton states these two laws as five Axioms, or self-evident truths, at the beginning of *Opticks*.

Fifty years earlier, Pierre de Fermat had enunciated a law stating that light always travels along the path of least time. Application of Fermat's principle (as it is generally known) leads directly to the laws of reflection and refraction when the relationship between the refractive index and the speed of light is taken into account.[9]

Application of the laws of reflection and refraction facilitates the design of modern optical instruments. An approximation in which the rays of light are assumed to pass close to the axis of the lens is often used to simplify the

---

[8]Strictly speaking it is the vacuum that has a refractive index of unity.

[9]As we shall see in the next Section, Fermat's Principle was used as an analogy by Lagrange in developing his *Principle of Least Action*, which led to a new formulation of mechanics.

mathematics required for the calculation of the paths of rays through lenses, and results in a simple formula for the location of the image formed by a lens. In cases where the ray passes near the edges of the lens, the situation is more complex, and the production of sharp images from such rays stretches even modern technology. Photography enthusiasts will be aware that images produced using the higher *stop numbers*[10] of their camera lens (i.e. smaller apertures, as the diameter of the aperture is inversely related to the stop number f) are generally sharper than those taken using the lens at its widest aperture.[11] Lenses with larger apertures are considerably more expensive than those with a limited aperture range because of the more complex design required.

4.6   Physical Optics: The Wave Theory of Light

Useful as the ray approach was to the description of a vast amount of physical data, it became clear by the 19th Century that there were optical phenomena that could not be explained by the corpuscular theory of light. Probably the definitive experiment was performed in 1803 by Thomas Young, who allowed coherent light[12] to fall on two closely spaced slits. When the light passing through these two slits was allowed to fall on a screen,[13] a pattern of bright and dark fringes was observed. This "interference" is a characteristic that occurs whenever two *wave trains*[14] pass through each other.

Interference effects are easily observed on a pond when two pebbles are dropped into the water. A set of concentric waves propagates out from the points of entry of the pebbles, and as the wave trains pass over each other there are some places where the waves are in phase and reinforce, and other places where they are out of phase and cancel out. If there are obstacles in the pond, such as a rock or post, an observer will notice that the ripples from a dropped pebble diffract around these to some extent and continue into the shadow region behind the obstacles. The waves may also rebound or scatter away from the obstacle, and these scattered waves may

---

[10]The stop or focal number (or simply f-number) is defined as the ratio of the lens's focal length to the diameter of the entrance pupil.

[11]There is a limit to this statement, because at the smallest apertures diffraction effects (see later) can cause blurring.

[12]When light is produced by a normal light source it is "incoherent" because the light comes from many unrelated atomic processes, and there is no phase relationship between these processes. As a consequence none of the interference effects we are describing will be observed. Young allowed the light from his source to pass through a small slit. This constriction reduced the number of light-emitting centres and made the light from the slit coherent enough for interference effects to be observed. Today coherent light of much greater intensity is produced by a laser.

[13]We are using the term "screen" loosely here. Because of the small separation of the fringes, they are usually observed through a microscope, and the "screen" is the retina of the observer's eye.

[14]A wave train is defined as a series of waves travelling in the same direction and spaced at regular intervals.

interfere with the incoming wave train. All of these phenomena have been observed with light, and by the 19th Century their cumulative effect was to put an end to the corpuscular theory of Newton in favour of a wave theory of light.

This is not to suggest that geometrical optics was now destined to become an unimportant curiosity to be studied only by those with an interest in the history of science. All the results of geometrical optics can be obtained from the wave theory in regions where the small-scale interference effects are unimportant, and the formulas of geometrical optics are much easier to use.

Even in Newton's time, interference effects were not unknown. The rainbow coloured fringes observed on the road in rainy weather when a drop of oil is smeared on the wet road surface is an example of interference. The oil has a different refractive index from water, and the light waves that are reflected from the air-oil interface and the oil-water interface interfere. White light is a mixture of different colours with slightly different refractive indices in oil. As a consequence, a region of positive reinforcement for one colour may coincide with a region of cancellation for another colour, producing the familiar rainbow patterns. This phenomenon was observed by both Robert Hooke and Isaac Newton, but is known today, somewhat unfairly given the history between the two men, as "Newton's Rings".

So, if it had now been firmly established that light is propagated as a train of travelling waves, the next question to ask was surely: "through what medium are the waves travelling?" In the example of surface ripples on the pond, it is obvious that without the water there would be no waves. Similarly, sound waves require a medium, usually air, for their propagation. Despite the multitude of loud whizzes and bangs film-goers are confronted with in sci-fi movies such as *Star Wars*, no sound is actually propagated through the vacuum of space.

Up until the middle of the 19th Century, physicists generally believed that there must exist a medium pervading throughout the universe to facilitate the propagation of light waves. This medium was called the aether (or ether). However, as we shall see in Chaps. 6 and 7, experiments to determine the velocity of light with respect to the ether failed, and the concept of an ether fell from favour.

Two main types of wave motion exist. The first is longitudinal wave motion, where the direction of the wave oscillations is backwards and forwards along the direction of propagation of the wave. Two examples of longitudinal waves are those passing along a coiled spring, and sound waves in a gas. The second type of wave propagation is transverse wave motion, where the direction of the oscillation is perpendicular to the direction of propagation. As there are two directions perpendicular to the direction of propagation, two differently polarised waves are possible.[15] These polarised waves do not interfere. Left and right-handed circularly

---

[15]If we consider the direction of propagation to be z in an x, y, z coordinate system, the oscillations can take place in the x or y direction, corresponding to horizontal and vertically polarised waves. Oscillations at other angles can be resolved into components in the x and y directions.

polarised waves are also possible, where the direction of oscillation rotates around the axis of the direction of propagation in a clockwise or counter-clockwise direction.[16]

Light exhibits all these forms of polarisation, thus confirming that it is indeed a transverse wave motion. Polaroid sunglasses are constructed from a material with long aligned molecules that allow only light of one direction of polarisation to pass through, absorbing the other. Glare from sunlight reflecting from non-metallic surfaces is also polarised, and the orientation of the polarisation axis of the sunglasses is set to eliminate this glare. Tilting one's head to the side while observing these reflections will cause the glare to reappear. Light from regions of the sky away from the sun is also polarised, and again, tilting one's head to the side while wearing Polaroid sunglasses will reveal changing patterns of intensity.

Some crystalline materials are birefringent, which means that the refractive index in one direction along the crystal surface is different from the refractive index in the direction along the surface at right angles to the first. These crystals can be used to produce circularly polarised light.[17] One application of this technology is in the production of 3D movies, where the 3D glasses worn by the viewer allow the left and right eyes to receive clockwise and counter-clockwise polarised images from the screen. These two images are slightly different and are combined by the brain of the viewer into a 3D image. One could as well have employed horizontal and vertical linearly polarised waves. However, in this case tilting one's head to the side, a prevalent occurrence during some of the activities found in the back rows of movie theatres, would destroy the 3D effect sought after so earnestly by the film director.

Towards the end of the 19th Century it must have seemed to physicists that Newton's corpuscular theory of light was dead and buried. However we will see in Chap. 6 that all was not quite as it seemed at the time. The corpse of Newton's brainchild was already stirring again, and a comeback to rival that of Lazarus was underway and gathering strength.

4.7 Beyond Newton—Analytical Mechanics

We saw in the first section of this chapter that Newtonian mechanics describes the motion of bodies in the world about us with an accuracy that is unquestioned even today. We impose a caveat that we are excluding from consideration very small

---

[16]Circularly polarised waves are produced when both horizontal and vertical polarisations are present but are one quarter of a cycle (i.e. 90°) out of phase with each other.

[17]The different refractive indices mean that the velocity of light is slightly different for oscillations in the x and y directions. The y wave therefore falls behind the x wave as the light propagates in the z direction. By choosing the thickness of the crystal appropriately one can ensure that as the light emerges from the crystal the y oscillation is one quarter of a cycle behind the x oscillation. The emerging light is then circularly polarised.

objects (atomic size and smaller) and objects travelling at speeds approaching that of light. These are domains that were certainly beyond Newton's ken and are outside of our everyday experience.[18] The speed of light is approximately 300,000 kilometres per second, about a million times the speed of a subsonic jet airliner. So, if Newtonian mechanics can explain it all, why did later scientists, in particular Joseph-Louis Lagrange and William Rowan Hamilton in the late 18th and early 19th centuries, develop a formulation of mechanics that is quite different from Newton's?

Both men were gifted mathematicians and probably needed little motivation for their work other than the intellectual challenge. However Newtonian mechanics, although easy to apply when the driving force is constant (i.e. not varying in space) can present considerable computational difficulties in other scenarios. This is not to say that the Lagrange/Hamilton formulation is easy—it isn't and makes use of a branch of mathematics called the *Calculus of Variations*, which is beyond the scope of this book. However, for some applications it provides a solution that would be very difficult to obtain using the three laws of Newton. We have already encountered one example, the rotating carousel, in the first section.

Does the new approach supersede that of Newton? Was Newton, in some sense, wrong? Not at all. Newton's laws can be derived from the Lagrange/Hamilton formulation, so everything that Newton predicts is also predicted by the later approach, which is often called *Analytical Mechanics* to distinguish it from *Newtonian Mechanics*. We are going in the following to give a brief description of Analytical Mechanics, warning on one side that our treatment is by necessity very cursory, and on the other that its abstract nature should not frighten the reader, since what is more relevant here is just to get the flavour of it.

4.8   The Method of Lagrange

Let us consider the simple case of a projectile fired with an initial velocity **u** at an angle θ to the horizontal, as shown in Fig. 4.2. From Newtonian mechanics we can calculate the path of the projectile, a parabola, as it falls under the influence of gravity. This is easily done because gravity exerts a constant downward force on the projectile. We can calculate everything about the projectile (its velocity, the time taken to reach any point in its flight, the height reached, etc.) from an application of Newton's 1st and 2nd laws of motion. This problem is commonly set in entry-level physics courses. However, if we are looking at the case where a rocket is travelling away from the earth through gravitational fields that vary with its position in space, the problem of determining its trajectory becomes much more difficult.

---

[18]In Chaps. 7 and 8 we will discuss Relativity and Quantum Mechanics, which deal with these excluded domains.

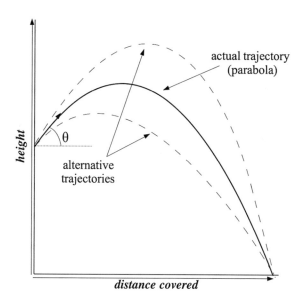

Fig. 4.2 Actual trajectory of a projectile fired with an initial velocity **u** at an angle θ to the horizontal (*solid curve*) can be easily obtained from Newtonian Mechanics. Possible alternative trajectories (*dashed curves*) are excluded in Analytical Mechanics by the Principle of Least Action (see text)

Returning to Fig. 4.2, let us consider the trajectory of the projectile once more. As we saw in Chap. 2, at every point along its path, the projectile has both a Kinetic Energy (KE) and a Potential Energy (PE). When it is first fired, it is fast and has a large KE. As it rises in its trajectory, this KE is partially converted to PE because of the work that the projectile has done against the opposing gravitational force. We saw how energy can be converted from one form to another in Chap. 2. As the object falls back to the ground this PE is converted back to KE.

Lagrangian mechanics is formulated around a function (the Lagrangian), knowledge of which essentially specifies the laws of physics for a system. In the particular cases we have been considering so far, where the forces involved are "conservative",[19] the Lagrangian can be defined as:

$$L = KE - PE.$$

As we see from Fig. 4.2, as the object travels along the trajectory, the KE and PE both vary in magnitude, and hence so does the Lagrangian. The sum (or integral in time) of the Lagrangian for every point along the trajectory is known as the Action, and is usually given the symbol A. Now where is this leading? Bear with us, for we are almost there.

The trajectory that we have shown in Fig. 4.2 was initially obtained from Newton's laws. However, putting that aside for the moment, we can imagine other possible trajectories that the projectile might follow. These are shown in Fig. 4.2 as

---

[19]Conservative forces are those which can be expressed, apart from a minus sign, as the gradient, or slope, of a potential field.

dashed lines. This leads us to the basic premise of Analytical Mechanics, namely the *Principle of Least Action*.[20] This principle states that of all possible trajectories that a projectile might follow, the one that nature selects for it is the one where the Action is at a turning point. (i.e. a minimum, maximum, or saddle point.)[21]

The application of the *Calculus of Variations* to determine the turning point of A leads to an equation known as the Euler-Lagrange equation, which is the basic equation of the Lagrangian formulation of classical mechanics.[22] If we know the Lagrangian for a system, we can solve the Euler-Lagrange equation to obtain the equations of motion of the system.

But we already knew in Fig. 4.2 which trajectory was the correct one, so what use is all of this? True, but as we have discussed earlier, there are other scenarios where Newton's laws are difficult to apply, e.g. on a carousel. Now Lagrange has given us an alternative technique for our use. We consider all possible trajectories and apply the Euler-Lagrange equation to find the one corresponding to stationary Action. If we were able to apply Newton's laws, we would get the same answer, for the two approaches can be shown to be completely consistent with each other. For some problems Newton's laws provide the quickest solution, but in others it is much easier to use the Lagrangian method.

There is an inherent intuitive appeal to the Newtonian approach. It solves a physical problem by commencing with a knowledge of the initial state of the system and calculates the effect of the applied forces. It is a direct approach. In comparison, Lagrange's method is much more abstruse. Imagine the reaction of a car driver who asks his navigator for the shortest route from A to B, only to be told to try all possible routes and select the one that uses the least fuel. Yet this is precisely what Lagrange does. How does nature know which trajectory minimises the Action? We don't know. Lagrangian mechanics correctly predicts the results of experiment, and for physicists that is usually sufficient.

As is often the case, a new way of approaching a physical problem can lead to new insights. In some cases, forms of symmetry can be seen in the Lagrangian. Whenever this occurs it has profound implications and indicates the presence of a conservation law.[23] For instance, a system for which the Lagrangian is unchanged when the system is translated in spatial position satisfies the law of conservation of momentum. But surely all systems satisfy the law of conservation of momentum? Actually, no. Only if the force applied to the system is "conservative" (see earlier). For example, the force of friction is not conservative, and when it interacts with a system, momentum is not conserved. This is evidenced by the slowing down of a

---

[20]This term is actually a misnomer, and it would be better called the *Principle of Stationary Action*. In calculus, when the derivative of a function with respect to a variable is zero, the function is at a turning point, which may be either a minimum, a maximum, or a saddle point.

[21]The *Principle of Least Action* has an earlier analogue in optics called *Fermat''s Principle*, which we discussed in the last Section.

[22]It is a second order differential equation, or alternatively two simultaneous first order differential equations.

[23]The relation between symmetry and conserved quantities is known as Noether's theorem.

rolling ball. However, if one could take into account the interaction of the atoms of the ball with the surface it is rolling on, and the interaction of these atoms with those of the whole earth, then yes, momentum would be conserved. Normally, however, for simplicity friction is applied to a problem as a non-conservative force.

Another example of symmetry is when the force acting between two objects depends only on their radial separation.[24] In this case the resultant Lagrangian is unchanged under a rotation of the axis. This symmetry is an indication that the angular momentum of this system is conserved.

4.9   Hamilton's Approach

A further extension to the Lagrangian approach was formulated by Hamilton. He defined a function, which today is known as a Hamiltonian, by combining the Lagrangian with an extra term. In the simple case of the projectile that we have been considering, the Hamiltonian H can be written as:

$$H = KE + PE$$

and can be seen as an expression for the total energy of the system. This formulation leads to two simultaneous first-order differential equations, which must be solved to obtain the equations of motion of the system.

By now the reader may be wondering whether mathematical physicists have lost the plot, and are introducing complications for their own sake. Surely we could have stayed with Newton. We have already discussed a number of instances where the Lagrangian is an easier method of solution for some problems. However, it is more far-reaching than that. The Lagrangian approach is much more general than Newton's and can be applied in domains where Newton's is out of the question, e.g. the Standard Model of Particle Physics (see Chap. 10). Newtonian mechanics is a special case of Lagrangian mechanics; they are not completely equivalent.

Then what of the Hamiltonian? Possibly Hamilton's ideas would have remained a curious academic backwater if it were not for one thing: as we will see in Chap. 8, the science of Quantum Mechanics is formulated in terms of the Hamiltonian. The interactions of atoms, electrons, protons, neutrons and other fundamental particles are expressed in terms of a quantum version of Hamilton's function. An understanding of the Hamiltonian is essential in the mechanics that is used to explain so much of twentieth century physics. However, we will leave further discussion to the relevant later chapters.

We conclude this section with an observation. The logical approach to mechanics would be to commence with a formulation of quantum mechanics, and work backwards, deducing classical mechanics as a special case of quantum

---

[24]This is usually called a "central force problem".

mechanics when the energies involved are large. However, this is not the way that physics usually evolves in practice. It tends to proceed from the simpler "special cases" to the more general. It may be several hundred years before sufficient understanding has been acquired to place the whole of a field in perspective. Perhaps we should remember this when we come to discuss some of the developments that are currently underway in the 21st century at the frontiers of knowledge.

In the next chapter we investigate systems comprised of many interacting bodies. It is unusual in the world around us to find small isolated systems. Usually there are many bodies undergoing mutual interactions, e.g. air molecules in a glass bottle, or water molecules in a droplet. Understanding the physics of such systems might appear intractable, but 19th Century physicists discovered ingenious approaches that are still in use today, and which form the topic of the next chapter.

References

1. A. Clarke, Hazards of prophecy: The failure of imagination in the collection. *Profiles of the Future: An Enquiry into the Limits of the Possible* (1962, rev. 1973), pp. 14, 21, 36
2. Isaac Newton, letter to Robert Hooke dated 5 Feb 1676

Chapter 5
Statistical Mechanics and Thermodynamics

Heat won't pass from a cooler to a hotter.
You can try it if you like but you'd far better not-a.
'Cos the cold in the cooler will get hotter as a rule-a.
'Cos the hotter body's heat will pass to the cooler ....
Michael Flanders and Donald Swann, *At the Drop of Another Hat*

Abstract The difficulty of dealing with the interactions in a many-body system are explained, and the origins of the statistical approach to this problem outlined. This leads into the statistics of gas molecules and the Maxwell-Boltzmann distribution of molecular velocities. The laws of thermodynamics are introduced, and the concept of entropy and the arrow of time are presented.

5.1 Many Bodies Make Light Work

We have seen earlier in the last chapter how the methods of Newton, Lagrange and Hamilton can be used to obtain the equations of motion of an object acting under the influence of a force. The force might be produced, for instance, by the gravitational or electromagnetic interaction with another object.

Now suppose we introduce a third object to complicate the scenario. If we label the objects A, B and C, then we now have a situation where A, B, and C are mutually interacting. We cannot consider in isolation the motion of A produced by the presence of B because at the same time A and B are both being influenced by C. A classic example of such a problem, which dates back to Newton's *Principia*, is the calculation of the orbits of the moon about the earth and their combined motion about the sun.

This three body problem attracted the interest of many physicists over the centuries, and is usually tackled by a method of successive approximations. In 1887 Ernst Bruns and Henri Poincaré showed that no general analytic solution exists for the three body problem.

If understanding the interaction of three bodies presents such difficulties, what hope have we of ever unravelling the motion of the billions upon billions of molecules in a small flask of gas? Fortunately, there is strength in numbers, and

© Springer International Publishing Switzerland 2016
R. Barrett et al., *Physics: The Ultimate Adventure*, Undergraduate Lecture
Notes in Physics, DOI 10.1007/978-3-319-31691-8_5

although the motion of an individual molecule is not accessible, their combined properties *en masse* lend themselves to study through the use of statistical techniques. This approach was coined *Statistical Mechanics* in 1884 by American Physicist, J. Willard Gibbs.

We are all familiar with the concepts of the temperature and pressure of a gas in a container. However, if we examine only a few molecules, temperature and pressure have little meaning, since they both arise as a consequence of the combined motion of all the molecules in the gas. Does this mean that we should write down Newton's laws for these molecules and attempt to predict how they interact? Obviously such an approach is not feasible.

Instead we think of average quantities. The temperature is related to the average kinetic energy of the molecules and the pressure to the average impact of the molecules on the walls of the container. (Of course, these walls are themselves made up of molecules, but they are tightly bound into a solid, and for our purposes we may consider the walls as a continuous rigid entity.)

The reader may feel it is unsatisfying to deal with averages in this way. We have all heard people talk of "*the law of averages*", by which they mean that because the coin they are flipping has come down heads four times in a row, the next toss must almost surely bring a tails. However, there is no such a law. If the coin is fair and unbiased, each throw is an independent event, and the probability of a tails at the next throw is the same as that of a heads, i.e. 50 %. But if the coin is flipped a large number of times, the number of heads and tails achieved will invariably be nearly equal.

It is worth pursuing this example a little further. In Chap. 3 we saw that an experimental measurement has an associated error. In the above example the error can be calculated from the laws of statistics. The average number of times a heads will result from a series of n throws is $n/2$, and the standard deviation[1] associated with this average (see Fig. 1, Chap. 3) is $\frac{\sqrt{n}}{2}$. After n flips of the coin we expect $n/2 \pm \frac{\sqrt{n}}{2}$ heads to turn up, which corresponds to an expected relative uncertainty of $\frac{1}{\sqrt{n}}$. Therefore if we flip the coin 100 times, we can anticipate a rather high error (10 %) in our prediction of 50 % heads being achieved. However, if we flip the coin 10,000 times the error in our 50 % prediction of heads drops to 1 %, and a million flips should reduce the error to 0.1 %, always assuming that the coin is still

---

[1]Those with an interest in statistics will realise that flipping a coin results in a binomial probability distribution with the probability of k heads being observed in n throws given by

$$P(k) = \{n!/[k!(n-k)!]\} * n^k * (1-n)^{n-k}$$

where $\pi = 1/2$ for an unbiased coin. The mean of this distribution is n/2 and the variance is n/4. When n is large, the binomial distribution can be approximated by a Gaussian distribution with the same mean and variance. The standard deviation of a Gaussian distribution is the square root of its variance, which leads to the result we have used.

unbiased at this level of accuracy, and that our flipping method does not introduce some unknown form of bias.

If the statistical accuracy seems quite good with a million flips of a coin, bear in mind that the number of molecules interacting inside our container of gas is astronomically large (of the order of 10 with 22 trailing zeroes[2]). Any conclusion based on the average properties of gas molecules will be accurate to a very high degree indeed. It is this principle which gives the power to statistical mechanics and makes it a very worthwhile and productive field of study in physics.

5.2 Kinetic Energy of Gas Molecules

Probably the earliest outcome from the application of this statistical approach to physics was the *Kinetic Theory of Gases*. The theory begins with a few basic assumptions, namely that a gas is composed of identical molecules which are in a state of random motion. The molecules are widely separated, obey Newton's laws of motion, and undergo collisions with each other that are perfectly elastic. The latter statement means that the molecules do not lose any kinetic energy by excitation of internal states during the collision with other molecules.

If we understand that the temperature of the gas is related to the average kinetic energy of the molecules, it should be clear that increasing the temperature of the gas, by increasing the velocity with which the molecules impinge on the wall of the container, will increase the pressure they exert on the wall. Following this type of argument through to completion leads to the *Ideal Gas Law*, which expresses the relationship between the temperature, pressure and volume of a gas in the form:

$$PV = kT$$

where P, V and T are the pressure, volume and temperature of the gas respectively, and k is a constant.[3]

Probably the first example of what became known as "*Statistical Mechanics*" was the derivation of the probability distribution of the speed of atoms in a gas by Rudolf Clausius in 1857. His initial work was quickly progressed by James Clerk Maxwell. Ludwig Boltzmann generalised their ideas to obtain the energy distribution of the molecules in a more rigorous fashion. Somewhat unfairly, the resultant distribution became known as the Maxwell-Boltzmann distribution. In his derivation, Boltzmann used the assumptions that we have listed above, but also assumed that the molecules were distinguishable.

---

[2]In chemistry it is known that there are Avogadro's Number (i.e. 6.022×10^{23}) of molecules in one mole (about 22.4 L) of gas at Standard Temperature and Pressure. One litre of air would therefore be expected to contain approximately 2.7×10^{22} mol.

[3]k is normally expressed as k = n R where n is the number of moles of the gas present and R is the Universal Gas Constant.

Fig. 5.1 Example of a
Maxwell-Boltzmann
distribution, showing the
number of particles per unit of
speed versus molecular speed
at different values of the
temperature in degrees Kelvin
(To convert degrees Celsius to
degrees Kelvin, add 273.15.
This conversion is explained
in the next section). The area
under each curve is the same
and is proportional to the total
number of molecules in the
gas

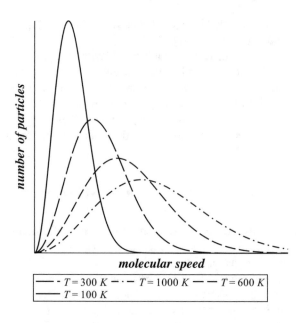

The form of the Maxwell-Boltzmann distribution is shown in Fig. 5.1. As the temperature is increased, the average velocity of the molecules increases, and the distribution skews towards higher velocities, as expected.

The middle of the nineteenth Century, the time when the ideas of statistical mechanics were evolving, was a period of industrial innovation in Europe. The advent of steam engines in factories, railways and ships, together with new methods for large scale production of steel, saw a burgeoning of industrial activity. A desire to increase the power and efficiency of steam engines was a motivation for the birth of a new branch of physics, which became known as *Thermodynamics*.

5.3   Entropy and the Laws of Thermodynamics

Thermodynamics began with a classical approach to the thermal and mechanical properties of macroscopic systems, with no consideration given to the underlying microscopic models. It dealt with heat transfers and work outputs from systems, and was later extended to treat chemical processes.

The keystone of thermodynamics is a set of laws, the first of which is the now familiar *law of conservation of energy*, which states (as already mentioned in Chap. 2) that energy cannot be destroyed, but only changed from one form to another. The total energy in the universe therefore remains constant. This statement assumes of course that we know precisely what energy is.

The second law (also mentioned in Chap. 2) is more subtle, and is considered by some to be the most profound law of classical physics. It has been stated in several forms, the simplest of which is probably: *"heat cannot spontaneously flow from a colder to a hotter location."* Consider two flasks of gas, one at a higher temperature than the other. If we connect the two flasks with an inter-joining tube, heat is gradually transferred from the hot flask to the cold one, stopping when the gas in both flasks is at the same temperature. We never have a situation where heat flows in the other direction, thereby increasing the temperature differential between the gases in the two flasks.

Another way of looking at the second law is to consider the concept of order. If by *order* we mean that we have more of a separation of objects of different type, we may state that the initial situation with hot gas and cold gas neatly separated into different flasks is more *ordered* than the final state, when the gas is completely mixed up. Left to itself, nature tends to the state of maximum disorder. The universe is analogous to a child's playroom, where the toys start out in the morning neatly arranged on shelves and in boxes, but at the end of the day have become strewn randomly over every horizontal surface. The term *"entropy"* has been coined for a measure of this disorder, and another expression for the second law states that *"entropy tends to a maximum"*. In the example of our gas flasks, maximum entropy has been reached when the gas is at a uniform temperature throughout. In this case it is said to have reached thermal equilibrium. The concept of entropy pervades much of modern physics.

The second law of thermodynamics received some notoriety in the 1960s when British scientist and acclaimed novelist, C.P. Snow, disparaged the lack of basic scientific knowledge among those who had received a classical education. He opined that ignoring the second law of thermodynamics was an equivalent in ignorance to never having read Shakespeare. Two satirists, Michael Flanders and Donald Swann, resolving to remedy the gap in their education, studied the law and based a song about it in their stage show and a later recording.[4]

Leaving the second law for the moment, we move on to the third law, which states: *it is impossible to reach the absolute zero temperature by any finite number of processes.* As an object is cooled, the kinetic energy of the molecules is reduced, until at the absolute zero of temperature we would expect them to become motionless.[5] The absolute zero of temperature is found to lie at $-273.15\ °C$, which is defined to be the zero in a new "Kelvin" temperature scale. The third law states that 0 K can never be attained, but only approached as the limit of a converging process of cooling.

Thermodynamics proved very useful in understanding and improving *"heat engines"*. Although the second law does not allow heat of its own accord to flow from a cold region of a system to a hot one, such a flow is possible if work is done

---

[4]Michael Flanders and Donald Swann *"At the Drop of Another Hat"*. See citation at the start of this chapter.

[5]We will see in a later chapter, that even at 0 K the molecules retain a *"zero point energy"*.

on the system by an outside agency. This is the principle behind refrigerators and
air-conditioners.

In 1824 Nicolas Leonard Sadi Carnot began the analysis of an idealised heat
engine. His work, later extended by others, involved the cycling of a system
through a series of compressions and expansions. For part of the cycle, the system
is kept at constant temperature and for another part the process is adiabatic, which
means that no heat is allowed in or out of the system. The end result is that the cycle
can be used to effectively separate the system into hotter and cooler components,
but only at the cost of performing work from an external power source.

Some readers may have puzzled over a statement, often made by air-conditioner
merchants, that heating a room with a reverse-cycle air-conditioner is cheaper than
heating it directly with an electric heater. Surely both methods convert electricity
into heat, and shouldn't the cost therefore be the same, all other considerations
being equal?

Well, actually no. The difference can perhaps be illustrated by imagining that we
have at our disposal two little imps. The first is equipped with a cricket bat (or
baseball bat, if you are a follower of that sport), and when we switch on the electric
power, the little fellow runs berserk through the room, smiting every molecule it
sees, goading them all to greater activity. As we know, faster molecules cause a rise
in temperature, and so the room gets hotter. This is what happens when we switch
on an electric heater.

When we switch on an air-conditioner, the second imp ambles over to an
opening leading out of the room, where it has previously installed a door. Whenever
a slow moving molecule approaches and is about to bounce off the door, it throws
open the portal, allowing the molecule to escape from the room. The average
velocity of the remaining molecules is thereby increased, and the temperature of the
room rises (of course some external reasonably warm molecules must be let into
compensate for the loss). It requires less work to operate the door than to run around
with the bat, hence the air-conditioner is more efficient at heating the room than an
electric fire. The air-conditioner is an example of a *heat pump* and its *Coefficient of
Performance* (COP) is defined as:

$$COP = \frac{work\ supplied\ to\ or\ removed\ from\ the\ room}{work\ consumed\ by\ the\ heat\ pump}.$$

A COP value of about 4 is typical,[6] and is indicative of the efficacy of the second
imp in the example above, compared with the first, whose work can never be less
than the energy gained by the system, and whose COP is therefore less than 1.

The efficiency of the heat engine studied by Carnot can also be calculated and is
found to be:

---

[6]There are legal Minimum Energy Performance Standards (MEPS) in place for air conditioners in
many countries. In Australia the COP standard for non-ducted split systems is 3.66.

$$Efficiency = \frac{work\ done\ by\ system}{energy\ put\ into\ system}$$
$$= 1 - \frac{temperature\ of\ cold\ reservoir}{temperature\ of\ hot\ reservoir}$$

The temperatures are measured in the Kelvin scale. For a steam engine, the cold reservoir is the external atmosphere and the hot reservoir is the steam boiler. The boiler is a pressure vessel, enabling the steam to become superheated, thereby increasing the efficiency of the engine. The efficiency given by the above formula is that of an ideal heat engine. All practical engines have a lesser efficiency.

With the development of statistical mechanics, it was realised that many of the classical components of thermodynamics lend themselves to an underlying statistical interpretation, and the term "*Statistical Thermodynamics*" has been coined for the resulting discipline.

The concept of entropy as a measure of disorder plays a leading role in statistical thermodynamics. In an isolated system, entropy tends to increase, but it may never decrease. This phenomenon has been dubbed the "*arrow of time*" (see Chap. 2) because observations of the level of disorder of a system at two different times enable an observer to determine which observation was taken first. However, surely here we have uncovered a paradox. As we stated earlier in this chapter, the laws of physics are reversible, and yet the Second Law presents a contrary example, where physical processes in an isolated system must evolve so that it proceeds inexorably to a state of increased disorder.

Some insight into this dilemma can be obtained by considering a room isolated from its surroundings so that no heat passes through its walls. Imagine the room contains only two molecules. What is the probability that both molecules are in the same half of the room?[7] Since each molecule has the same probability of being in the left or right half of the room, there are four possible *configurations*, each with a probability of 25 %. Two of those have both molecules in the same half (right or left) with a consequent total probability of 50 %. When the molecules have passed together into the same half of the room, the entropy has decreased, as this is a more ordered (less random) state of the system. The second law is thus invalid for 50 % of the time.

Suppose we now have three molecules in the room. What is the probability that they are all together in one half of the room? The earlier figure now drops to 25 %, and in the case of four molecules it drops further to 12.5 %. In a room containing N molecules, the probability that they are all in one half of the room is $\frac{1}{2^{N-1}}$. Given that N is approximately 10^{27} for a reasonable sized room, and that N appears as an exponent in the above expression, we can easily see why the Second Law represents the physical reality so extremely well.

---

[7]Here we follow the lead of Boltzmann and consider the molecules to be distinguishable. In Quantum Mechanics, the particles are identical, which leads to a different statistics, but the classical approach is adequate for our purposes here.

The type of statistical argument that we have employed above is widely used in Statistical Thermodynamics. It has produced an understanding of puzzling physical phenomena, such as ferromagnetism and various phase transitions that are difficult to explain with the conventional physics of Newton and Lagrange.

In the everyday world we encounter many examples where the entropy in our surroundings decreases. These may be as simple as when a parent tidies up a child's toys at the end of the day, or as complex as the evolution of life. Are these not examples of the violation of the Second Law?

No. We must not forget that the Second Law applies to an *isolated* system, i.e. one which has no energy input from an external source. The earth, however, is subjected to a continuous stream of energy from the sun. It is this energy that enables local processes to produce ordered systems on earth, from the annual blossoming of spring flowers to the newborn baby. If we include the sun in our isolated system, the net change in entropy is positive, as the Second Law predicts.

When we look at the universe as a whole, we observe a highly ordered system. Matter is arranged in galaxies of stars, many of which have planets orbiting about them, and on at least one planet life has evolved.[8] How is it that the universe is still so highly ordered after billions of years of evolution? If the Second Law is valid, shouldn't the cosmos by now consist of a more or less uniform distribution of particles? This is what one might normally expect in the aftermath from a massive explosion. This final state is the predicted "*heat death*" of the universe, where it has attained the state of maximum entropy and there are no remaining temperature differentials to drive physical processes. Why the universe still remains so ordered today is a question we will defer to Chap. 11.

In the next chapter we address one of the crowning achievements of nineteenth century physics, the theory of electromagnetism, as proposed by Scottish physicist, James Clerk Maxwell. We also examine some of the unsettling new discoveries being made at that time that lay outside the ambit of known physics, and heralded the birth of modern physics.

Reference

1. R. Heller, Sci. Am. **312**(1) (2015)

---

[8]Planets more habitable than Earth may actually be common: see [1].

Chapter 6
Electromagnetism and Cracks in the Edifice of Classical Physics

And Maxwell said:

$$\oint \mathbf{E} \cdot d\mathbf{A} = \frac{q_{enc}}{\varepsilon_0}$$

$$\oint \mathbf{B} \cdot d\mathbf{A} = 0$$

$$\oint \mathbf{E} \cdot d\mathbf{s} = -\frac{d\Phi_B}{dt}$$

$$\oint \mathbf{B} \cdot d\mathbf{s} = \mu_0 \varepsilon_0 \frac{d\Phi_E}{dt} + \mu_0 i_{enc}$$

and there was light.

Old Physicist's joke.

Abstract Electricity, magnetism and their unification by Maxwell into the theory of electromagnetism are presented as the crowning achievements of nineteenth century physics. The growing unease arising because of conflicts between Maxwell's theory and classical mechanics, the unexplained black-body emission spectrum, the photoelectric effect and the null result of the Michelson-Morley experiments, is a harbinger of the revolution in physics that took place in the twentieth century.

6.1 Electricity and Magnetism

In this chapter we address one of the crowning achievements of nineteenth century physics, the theory of *electromagnetism*, as proposed by Scottish physicist, James Clerk Maxwell. It was probably the success of Maxwell's theory that led Lord Kelvin and others near the end of the century to their somewhat premature conclusion that all of worth in physics had now been achieved. As we shall see later in the chapter, Maxwell's theory and other concurrent discoveries actually marked a turning point in the development of physics.

The set of equations describing electromagnetism put forward by Maxwell (and cited at the head of this chapter) has all the attributes of a good theory that we

© Springer International Publishing Switzerland 2016
R. Barrett et al., *Physics: The Ultimate Adventure*, Undergraduate Lecture
Notes in Physics, DOI 10.1007/978-3-319-31691-8_6

discussed in Chap. 3. It enables the prediction of experimental results to a high precision; it also unifies electricity and magnetism, two areas of physics that originally had been considered disparate. Unfortunately for our purposes, the theory makes use of vector calculus, a branch of higher mathematics that is beyond the level at which we wish to pitch this book. The difficulty lies more in the integrals than in the vectors. However, we will attempt here to explore the underlying physics as far as possible within the limitations we have set.

The phenomenon of static electricity has been known since the time of the Ancient Greeks. In 600 BC Thales of Miletus was aware that the material amber, when rubbed with animal fur, attracts light objects. Even earlier, the Egyptians knew that some species of fish, when touched, deliver what is now recognised as an electric shock.

From the seventeenth century onward, European scientists added to the burgeoning knowledge of this strange phenomenon, which in 1600 William Gilbert had named "electricity" from the Greek word for amber. Machines to produce static electricity were invented, and it was discovered that electric forces could be transmitted through a vacuum. Positive and negative electric charges were identified, and like charges were found to repel each other and unlike charges attract.

In 1752 Benjamin Franklin, in a famous kite-flying experiment, demonstrated that lightning is a form of electricity. Since then several others trying to replicate the experiment have perished in the process, and the experiment, which involves flying a kite in a thunderstorm, is extremely dangerous.

In 1767 Joseph Priestley discovered that the electric force followed a similar inverse square law with the separation of the charges, as Newton had found for gravity. In 1800 Alessandro Volta invented the first electric battery and demonstrated that electricity could flow over wires. This last development heralded the harnessing of electricity for human exploitation.

Magnetism was another phenomenon that was first observed by the ancient Greeks, and the Chinese, in about 2000 BC. It was noticed that a material, which was given the name lodestone (literally "guiding stone"), had the ability to attract iron. Lodestone contains magnetite, a naturally occurring magnetic material, Fe_3O_4. Because of its strange attributes, lodestone was believed to possess magical properties. From 4500 years ago the Chinese exploited the tendency of lodestone to align itself in the north-south geographical direction when suspended, or floating in a liquid, to develop a compass for maritime purposes.

Apart from its use in compasses, lodestone remained for centuries little more than a scientific curiosity. It was noted that the earth itself was magnetised, and that artificial magnets could be made by aligning a piece of iron in the north-south direction and striking it with a hammer. Heating a magnetised material was an effective way to destroy its magnetic properties. Also, the attraction between magnets seemed to bear no relation to the attraction experienced by light objects when placed near a piece of amber that had been rubbed with fur. The magnetic forces appeared to emanate from each end of the magnet. These regions were dubbed the "poles" and were found to be of two different types, north and south, depending on whether they aligned towards the north or south when used as a compass. It was found that like poles of a magnet repel each other, whereas unlike poles attract.

In Chap. 4 we discussed the concept of gravity, and the enigma, unexplained by Newton, of the ability of the gravitational force to act at a distance across empty space. An object subjected to the gravitational pull from a second object is said to be under the influence of that object's gravitational field. Now, with electricity and magnetism, we have two additional phenomena which exhibit the puzzling action-at-a-distance that had so worried Descartes and Leibniz about Newton's theory of gravity. The situation is even more complicated because the gravitational force is always attractive, whereas the electrical and magnetic forces can also be repulsive.

A ground breaking development occurred in 1820 when Hans Christian Ørsted discovered serendipitously that an electric current passing through a wire deflected a nearby compass needle. This marked the beginning of the unification of electricity and magnetism, which hitherto had been considered two independent phenomena. The next ten years was a period of rapid progress. The electromagnet was developed, followed by the first electric motor invented by Michael Faraday. By 1837 the first industrial electric motors had been constructed.

Concurrently with the development of practical applications, the underlying physical laws were also coming to light. In 1826 Georg Simon Ohm discovered the law that bears his name:

$$V = I\,R$$

relating electric potential difference V, electric current I and the circuit resistance R. Joseph Henry and Michael Faraday discovered the principles of electromagnetic induction, power generation and transmission. The stage had been set for Maxwell's ground breaking work in 1865: the publication of his seminal paper, *A Dynamical Theory of the Electromagnetic Field*.

6.2 Maxwell Brings It Together

Before discussing Maxwell's achievements, it is necessary to clarify what physicists meant, and mean, by the concept "*field*".[1] A basic tenet of early natural philosophy, about which there was almost universal agreement, was that two bodies needed to be in contact with each other to interact. Action at a distance was not possible. Slowly, however, it became apparent that there were special cases when objects appeared to exhibit a sphere of influence on their surroundings. Electricity, magnetism and gravity were examples. However, the "*mechanical philosophers*", one of whom was Descartes, would not countenance this explanation. Even Newton, as

---

[1]The concept of *field* will be discussed further in Chaps. 8 and 10.

we have seen, was hesitant to embrace "*action at a distance*" fully, and left the question hanging.

Nevertheless by the middle of the nineteenth century, physicists had accepted the concept of action at a distance, and used the term "field" to describe the area of influence of an electric charge, magnet or gravitational mass. Faraday, after much intellectual wrestling, decided that his experiments on electric motors could best be understood by postulating inductive forces filling all of space around magnets. These "*lines of force*" emanated from the poles of the magnets, and can be rendered visible by placing a piece of card over a magnet and sprinkling the card with iron filings.

The term "field" had become commonplace enough by the mid-nineteenth century for Maxwell to use it in the title of his famous paper. To name something is to recognise its existence and inspire investigation into its properties. The properties of electric and magnetic fields became widely investigated. However, familiarity does not necessarily correspond to understanding in any deep sense, whatever that may mean. A physicist might argue that the accurate prediction of experimental results is the most that can be hoped for.

In modern times Descartes may have landed a blow from the grave, as fields in modern physical theories are considered to be produced by the exchange of particles, albeit virtual ones,[2] somewhat along the lines of the principle of contact action to which he so strongly adhered. We will discuss this concept further in Chap. 10.

The contribution of Maxwell was to recognise that oscillating electric and magnetic fields are related in a particular way. Any oscillating electric field will generate a corresponding oscillating magnetic field, and vice versa. It is not possible to have one oscillating field without the other. There is a symbiotic relationship between the two phenomena, so much so that they are now usually referred to jointly as *electromagnetism*. This type of symmetry, or dualism, is common in physics (see Chap. 2). So we have left-right symmetry, where the laws of physics in Alice's Looking Glass world are the same as in ours (except for processes involving the weak nuclear interaction, as we saw in Chap. 3); we have matter and antimatter, although one is much more prevalent than the other; we have positive and negative charges, north and south magnetic poles, and so on. These dualities often arise from symmetries in the underlying mathematical equations, and help scientists in the intuitive phase of their work, i.e. the formulation of hypotheses to be explored further. The symmetry between electricity and magnetism has resulted in many endeavours to discover a magnetic monopole, a single isolated magnetic pole analogous to an electric charge. So far these investigations have been unsuccessful.

The dualistic nature of the electromagnetic interaction becomes apparent if we generate, for example, an oscillating electric field by moving an electric charge to and fro. The oscillations of the electric field propagate through space as a wave, in the same manner that wiggling your hand up and down in a pond generates ripples

---

[2]Virtual particles exist for only a very brief period of time. They will be discussed later.

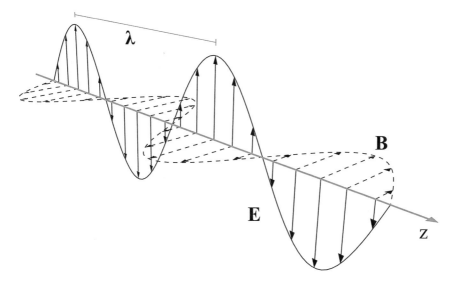

Fig. 6.1 Propagation of an electromagnetic wave in the positive z-direction. The curves labelled **E** and **B** show the electric and magnetic field strengths, respectively. λ is the wavelength of the wave

that move outward across the surface of the water. These oscillations of the electric field produce a corresponding oscillating magnetic field, which propagates in step with the electric field. The directions of oscillation of the two fields are at right angles to each other, and are transverse to the direction of motion of the waves. The same situation would result if we began by initially producing an oscillating magnetic field. The structure of the electromagnetic wave is illustrated in Fig. 6.1.[3]

The nature of electromagnetic radiation revealed in Fig. 6.1 is not the consequence of a hand-waving conjecture by Maxwell, but arises directly from the four equations, or laws, that he proposed and which are reported in the *Old Physicist's joke* at the beginning of this chapter.

The first two equations state that the divergences[4] of the electric and magnetic fields are zero in free space. This can be explained by analogy. Imagine a perforated cage in a flowing river. In any given time interval, the volume of water that has flowed into the cage is equal to the volume that has flowed out of it. If this were not the case we would have either an accumulation or a depletion of water in the region.

---

[3]This image has been slightly modified from the image uploaded by Lenny Wikidata to Wikimedia Commons and made available under the Creative Commons Attribution-Share Alike 3.0 Unported license.

   (https://upload.wikimedia.org/wikipedia/commons/thumb/d/da/Electromagnetic_wave.svg).

[4]The divergence comes from the Divergence Theorem (or Gauss's Theorem) in vector calculus. The theorem states that the net flow out of a region is equal to the sum of all sources minus the sum of all sinks in the region.

As water is incompressible (largely), this is not possible. Similarly, the number of lines of force in magnetic and electric fields that enter an empty volume in space is equal to the number leaving it. Here, the density of lines of force at a point in space is taken to be proportional to the strength of the magnetic or electric field at that point.

The exception is if there happen to be electric charges within the enclosed volume. In this case, the electric lines of force can originate or terminate on the charges, depending on whether the charge is positive or negative, and Maxwell gives an explicit formula describing this situation. The same would be true if magnetic monopoles were ever discovered.

It should be recognised that although these two laws appear as the first two of the four laws associated with Maxwell, their origin lies with Carl Friedrich Gauss, whose name is associated with much of the physics and mathematics of the nineteenth century. The third of Maxwell's laws is the law of magnetic induction. It gives the strength and direction of an electric field produced by a varying magnetic field. The faster the magnetic field changes, the larger is the generated electric field. This law can be traced back to Michael Faraday.

The final one of Maxwell's four laws addresses the complementary situation where a magnetic field is produced by a varying electric field. It completes the last link that binds the oscillating electric and magnetic fields together into an inseparable unity.

A characteristic of a good physical theory is that it explains phenomena in an area of study quite removed from the one that inspired the underlying law(s) in the first place. From Maxwell's equations it is possible to deduce the speed of electromagnetic radiation through space from two fundamental constants. These are ε_0, the electric permittivity of free space, and μ_0, the magnetic permeability of free space. The former relates the strength of the electrostatic force to electric charges and their separation. It is analogous to the gravitational constant G in Newton's formula for gravity. The latter is a measure of the strength of the magnetic field produced by a moving electric charge (or electric current) in a vacuum.[5]

The values of these two constants can be determined in laboratory experiments unconnected with electromagnetic radiation.[6] Maxwell's equations predict that the speed of electromagnetic radiation through space is given by

$$c = \frac{1}{\sqrt{\varepsilon_0 \mu_0}}.$$

---

[5]The same role as G in gravity is played by $1/(4\pi\varepsilon_0)$ for electric fields and $\mu_0/(4\pi)$ for magnetic fields.

[6]The magnetic permeability, now usually known as the magnetic constant, can be determined from Ampère's force law, which expresses the interaction force between two parallel wires bearing an electric current. The permittivity, or electric constant, was traditionally determined from Coulomb's inverse square law for the force between electric charges.

Substituting measured values for ε_0 and μ_0 produced a calculated result for c which was close to the measured value for the velocity of light.[7]

Maxwell recognised the closeness of the calculated and measured velocities, and from the coincidence asserted that light is a form of electromagnetic radiation. It was now clear that the fields of electricity, magnetism and optics were all aspects of the same physical phenomenology. In Fig. 6.2 we present the full range of the electromagnetic spectrum, which extends from radio waves at the lowest frequencies, through microwaves, infrared, visible and ultraviolet light and x-rays to gamma nuclear radiation at the highest frequencies[8] (smallest wavelengths).

The explanation of so many different phenomena by one overriding physical theory was a huge triumph for the power of scientific analysis, as pioneered by Galileo. Yet, even as physicists at the end of the nineteenth century were rubbing their hands together in an ecstasy of self-congratulation, inconsistencies and doubts were already beginning to arise, that eventually would propel physics forward into a new era of discoveries leading to concepts that almost defy credibility.

In the remainder of this chapter, we will guide the reader through some of these growing doubts, before launching in Chap. 7 into what is generally known as *The Golden Age of Physics*.

6.3   The Beginnings of Doubts

The first of the problems confronting physics at the end of the nineteenth century involves the two standout theories of mechanics, as formulated by Newton, Lagrange and Hamilton, and of electromagnetism, i.e. Maxwell's theory. As we have already discussed, mechanics was so encompassing that it explained not only the movement of objects subject to forces in everyday life, but also the motion of the planets around the sun. Maxwell's theory had unified and quantified our knowledge of electricity, magnetism and light and predicted the results of experiments in mechanics and electromagnetism to a very high precision. And yet, on close inspection, they were found to be mutually incompatible. They could not both be right.

One of the basic principles of Classical Mechanics is that experiments that are performed in different locations under the same conditions must yield the same results. In fact, we can extend and clarify this principle and state that experiments performed in locations that are at rest or moving in a uniform straight line with respect to each other, yield the same results. Only if one location is accelerating with respect to the other will the results differ.

---

[7]This relationship between c, ε_0 and μ_0 is now so well accepted that it is used to obtain the value of ε_0 from known values for c and μ_0. See Chap. 3 for a discussion of measurements of c.

[8]The frequency ν, wavelength λ, and velocity c of a wave are related by the formula $c = \nu\,\lambda$.

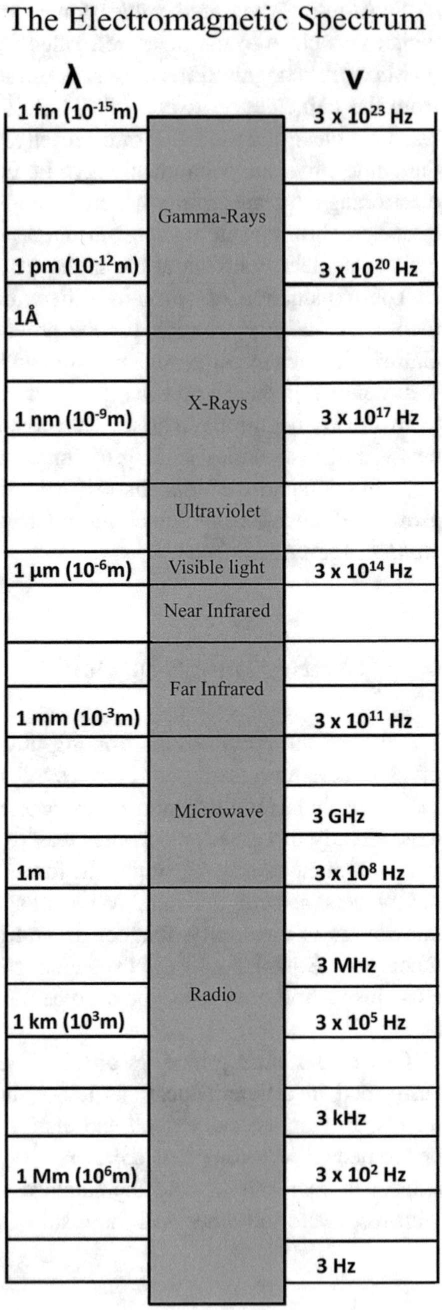

Fig. 6.2 The electromagnetic
spectrum showing the various
forms of electromagnetic
radiation as a function of
wavelength and frequency,
λ and ν respectively

The Electromagnetic Spectrum

λ		ν
1 fm (10^{-15}m)		**3 x 10^{23} Hz**
	Gamma-Rays	
1 pm (10^{-12}m)		**3 x 10^{20} Hz**
1Å		
	X-Rays	
1 nm (10^{-9}m)		**3 x 10^{17} Hz**
	Ultraviolet	
1 µm (10^{-6}m)	Visible light	**3 x 10^{14} Hz**
	Near Infrared	
	Far Infrared	
1 mm (10^{-3}m)		**3 x 10^{11} Hz**
	Microwave	**3 GHz**
1m		**3 x 10^8 Hz**
		3 MHz
	Radio	
1 km (10^3m)		**3 x 10^5 Hz**
		3 kHz
1 Mm (10^6m)		**3 x 10^2 Hz**
		3 Hz

When we are sitting in a smooth-flying aircraft we can relax and behave exactly as if we were sitting in our armchair on earth. Only at take-off, landing, and at moments of turbulence when acceleration effects become apparent, do we notice anything unusual.

In physics, this phenomenon is expressed by stating that the laws of physics are invariant under a Galilean transformation of the coordinate system. A Galilean transformation changes the physical laws, as expressed in one coordinate system, to the laws expressed in another coordinate system that is in uniform linear motion with respect to the first (e.g. from our lounge room on earth to the cabin of the jet airliner).[9]

The problem is that Maxwell's laws of electromagnetism are *not* invariant under a Galilean transformation. As if this were not enough, two experimenters, Albert Michelson and Edward Morley, were in 1887 about to contribute to the confusion.

If light is propagated by a wave motion, as Young's double slit experiment showed and Maxwell's theory explained, then the wave is expected to pass through some form of medium. Water waves cannot exist without water; sound waves cannot exist without air, or some other material through which the longitudinal compression waves of sound can pass. For light, the corresponding medium was dubbed the aether (or ether), and was assumed to pervade all space. Michelson and Morley reasoned that the ether must be in relative motion with respect to the earth (and therefore the laboratory) as a consequence of the earth's near circular orbit about the sun (and the sun's motion through the galaxy). Light would be expected to take a different time to travel a fixed distance up and down the ether stream compared with the time required to travel the same distance to and fro across the stream. Consider the analogy of an oarsman rowing a boat up and down a flowing river, and then from one side of the river to the other. The times taken will be different, even if the distances travelled are the same. Michelson and Morley set up an experiment to test this deduction.

Figure 6.3 shows the details of their experiment. Light from a source is collimated and passed on to a half-silvered mirror, which splits the light into two beams. The two beams are sent off along paths at right angles to each other. After travelling a certain distance, both beams are reflected back along their outgoing paths and allowed to intersect. Because they are coherent (both emanating from the same source), they produce an interference pattern of fringes in an analogous way to Young's earlier experiment. The whole apparatus was set up on a slab which floated on a pool of mercury. (This was a period in history when health and safety issues were not the concern they are today.)

---

[9]If we have two coordinate systems with common x, y and z axes, and the same origin at time $t = 0$, and one of the two (designated by the prime) is moving linearly with respect to the other with velocity v along the common x axis, the equations of motion in the unprimed coordinate system can be transformed into equations of motion in the primed coordinate system by the equations: $x' = x - vt$, $y' = y$, $z' = z$, and $t' = t$. The last equation shows that time is independent of the motion of the system in a Galilean transformation. This point will be discussed further in Chap. 7.

Fig. 6.3 The
Michelson-Morley
experiment. Light from a
source is split into two beams
which travel along paths at
right angles to each other,
before being recombined to
produce interference fringes

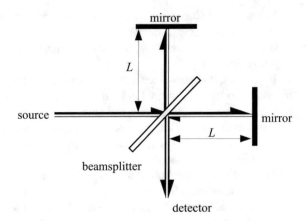

The anticipated result was that as the slab was rotated slowly through ninety degrees on the pool of mercury, the observed fringes would move across the screen. The rotation of the slab was expected to move the path of the first beam from one lying along the direction of the aether stream to one lying across it, and vice versa for the second beam. Because of the different times required to travel the two paths, the optical path difference between the two beams would change as the slab is rotated, which would give rise to the movement of the fringe pattern across the screen.

This was what was expected, but no change in the fringe pattern was observed. Michelson and Morley had produced what was to become arguably the most famous null result in the history of physics. The experiment essentially showed that the velocity of light was the same, irrespective of the velocity of the source and the receiver, surely an *impossibility*.

It required a young patent clerk working in Berne, Switzerland, to finally solve the two mysteries that we have outlined above. His name was Albert Einstein and he published his historic paper in 1905, giving birth to the *Theory of Relativity* and putting an end to the unchallenged reign of Classical Mechanics. We will discuss his *revolution* in Chap. 7.

6.4   Particles or Waves

The nineteenth century conception of the wave nature of light was also under attack on a number of other fronts. A problem arose when attempting to explain the spectrum of thermal radiation emitted by a black body. A black body is defined as an object that absorbs all radiation falling upon it. In the laboratory it is best constructed by excavating a cavity in an opaque material that is only partially reflective. A light ray entering the cavity will be reflected around inside it until it is absorbed. When heated, a black body gives off radiation, which is referred to as black-body radiation, or cavity radiation.

The emission spectrum of a black body when heated was a subject of study, both theoretical and experimental, in the nineteenth century. Normal objects, such as a bar of iron, when heated radiate an emission spectrum (emission intensity as a function of the frequency of the radiation) that approximates that of a black body. In one sense, the black-body can be regarded as the ideal emitter. Its spectrum is independent of the material used in its construction.

As the black body is heated, the peak of the emission spectrum moves with increasing temperature from infra-red, through red, yellow and eventually to blue-white. This phenomenon explains the varying emission spectra from different stars, and enables an estimate of a star's surface temperature.

However, all attempts to explain the spectrum of black-body radiation using classical statistical mechanics ended with a complete disagreement with experiment. Instead of the spectrum peaking and then falling away at high frequencies (or short wavelengths), as observed in the laboratory, the predicted spectrum continued to rise. This discrepancy was given a somewhat melodramatic name: the ultra-violet catastrophe. Figure 6.4 shows the experimental and calculated curves for the black-body radiation spectrum.

A theoretical explanation of the black-body spectrum was only achieved when a German physicist, Max Planck, introduced a completely radical and arbitrary assumption into the statistical theory, namely that the electromagnetic radiation from the cavity was not emitted continuously, but instead came out in discrete packets, or quanta. The energy of each packet is proportional to the frequency of the radiation; i.e.

$$E = h\nu$$

Fig. 6.4 Black Body Radiation Spectrum for several temperatures, compared with the classical theoretical prediction for 5000 K

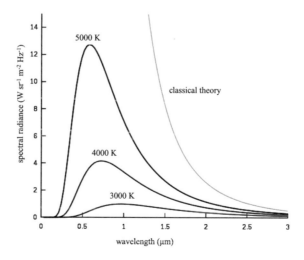

where E is the energy of each packet, ν is the frequency of the associated radiation and h is a constant, now known as Planck's constant.

Such a hypothesis may seem extreme to explain one anomalous physical phenomenon (remember Occam's Razor), but other disquieting experimental results had begun to accumulate. When light is allowed to fall on a metal surface (e.g. sodium) it was discovered that electrons are emitted. Energy from the light beam is transferred to electrons in the metal, until they accumulate sufficient energy to escape from the surface. An intense light beam, i.e. one where the amplitude of the electromagnetic oscillations is large, is expected to emit more energetic electrons than a less intense beam.

However, experiments on this photoelectric effect, as the phenomenon was dubbed, revealed nothing of the sort. It was found that the energy of the electrons is related only to the frequency of the light and not to its intensity. A more intense beam produces more electrons but not more energetic ones. Also to be noted is that light of a frequency below a certain threshold (which depends on the nature of the metal used) produces no photoelectrons, no matter how intense the beam of light. This experiment cannot be explained by Maxwell's wave theory of electromagnetism.

It was Einstein who in 1905 (a busy year for him) published the paper that won him a Nobel Prize, and truly gave birth to the quantum era. He extended Planck's ideas and proposed that a light beam was not a continuous wave, but rather was composed of discrete wave packets. These wave packets became known as photons. The energy of a photon is given by Planck's formula. When a photon falls on the metal surface, its energy is transferred to an electron, which is emitted from the surface with an energy E given by

$$E = h\nu - W$$

where W is the threshold energy which an electron must expend to overcome its bond with the metal, and is a characteristic of the nature of the metal.

So what have we here? Light behaves sometimes like a wave, as in Young's double slit and other interferometric experiments, and sometimes like a particle, the photon. The shade of Newton must surely have been enjoying a quiet chuckle at this confusion, which at least in part was a vindication of his particle theory of light.

Later on and after the onset of relativity and quantum mechanics, other experiments also found faults with Maxwell's theory. In 1922 light (or rather photons) was found to scatter inelastically from free electrons, a phenomenon known as the Compton Effect. The result was that the electrons gained kinetic energy, and the photons lost energy. This lost energy manifested itself as a reduction in frequency (or an increase in wavelength) of the scattered photons. The Compton Effect cannot be explained by Maxwell's wave theory, but is easily accounted for by the particle theory of light and the Planck-Einstein formula for the photon energy.

How could physics, which shortly before the end of the nineteenth Century seemed as neatly resolved as an Agatha Christie murder mystery in its last page, in the space of a few years be thrown so wide open with contradictions that every entry to the laboratory seemed to produce some new dilemma? Twentieth Century physics, with its array of new problems, attracted some of the greatest minds of the age. The solutions they proposed stretch human credibility even now. More than ever it has become necessary to rely on mathematical analysis, rather than intuition based on everyday experience.

Now that we have set the background, we are ready in the next chapters to commence our journey into the realm of what is loosely called "modern physics". It is time to take Alice's hand and step through the looking glass. What we encounter will at times seem to betray "common sense". We must cling to Occam's Razor to guide us along our way, and put our trust in human logic, rather than experience and old-fashioned common sense.

Chapter 7
Relativity

> There was a young lady named Bright,
> Whose speed was far faster than light.
> She set out one day
> In a relative way,
> And returned on the previous night.
>
> A.H. Reginald Buller

Abstract The chapter expounds the essentials of the theory of relativity, limiting as much as possible the use of mathematics, and trying to discuss some of the more astonishing consequences of the theory. The relativity of time is considered, as well as the implications of the generalization of dynamics and of the conservation laws to four dimensions. Following the approach of Einstein himself, the equivalence principle is used as the way to arrive at the general theory of relativity and to the inclusion of the gravitational interaction in the geometric properties of space-time.

7.1 A Bit of History

Einstein's relativity is, and has always been considered among non-scientists, the most glamorous theory of physics.[1] It was also thought to be so abstruse that only a handful of initiates in the world could understand it. What is so special about relativity that explains both its popularity and its halo of mystery?

Let us start with a bit of history. As we have seen in the last chapter, in the final decades of the 19th century, physics appeared to be a triumphant science. Physicists believed that the roots of all phenomena in nature had already been discovered and

---

[1]In fact by 1921 the Nobel Prize Commission was flooded by nominations for Albert Einstein. Since, however, the foundation statute required that prizes could be awarded only for research validated by an experimental confirmation (which was not yet the case for general relativity), the Commission was reluctant to yield, until a commissioner remarked that Einstein was *already* more famous than the Nobel Prize itself, so that conferring the prize to him would rather enhance the prize prestige, than vice versa. Then the *escamotage* was found to award the prize to Einstein *for his discovery of the law of the photoelectric effect*, without quoting relativity.

© Springer International Publishing Switzerland 2016
R. Barrett et al., *Physics: The Ultimate Adventure*, Undergraduate Lecture Notes in Physics, DOI 10.1007/978-3-319-31691-8_7

all that remained was to work out the details and contrive new applications for industry. These feelings were perfectly consonant with the dominant views of the time on economy and society, fully expressed (at least in the "West") by the ideology of "progress", i.e. the conviction that, despite social tensions and thanks to science and technology, humanity was destined to a bright future of ever increasing prosperity: "*le magnifiche sorti e progressive*" (*the magnificent and progressive destiny*), to which the Italian poet Leopardi alludes with a touch of bitter irony [1].

Leaving economy and politics aside, scientists believed that there remained only a few "*minor puzzles*" concerning physics in general, and in particular, the most refined intellectual achievement of the time, Maxwell's electromagnetism, which had crowned almost two centuries of continuous and progressive success. But what were these puzzles?

As we have already seen, Galileo had stated that the laws of physics must be the same for all inertial observers. This principle, known as *Galilean Relativity*, had received a consistent mathematical formulation with Newton. Who are these "*inertial observers*"? They are those who are isolated from the action of any force, so that they may only be in rectilinear uniform motion with respect to each other. The proclaimed invariance of the laws of physics for all inertial observers implied simple, reasonable and intuitive transformation rules for the coordinates used by different observers. These are the Galilean coordinate transformations, which were discussed briefly in Chap. 6. For the purposes of this chapter we now need to examine these transformations in more detail.

Let us consider a very simple example, with two observers setting up a reference system, in which the first one is located at the origin O of the coordinates, and the second one moves along the x axis, in the positive direction, with a constant speed V. Correspondingly, we consider an x' axis, superposed on the x axis, but with the origin O' moving with the second observer. This situation is illustrated in Fig. 7.1.

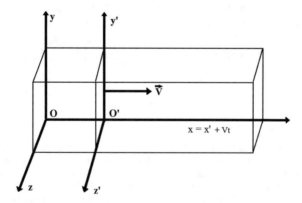

Fig. 7.1 Two Cartesian coordinate systems set up by two observers moving uniformly, with respect to each other, at the speed V along their x and x' axes: O' is moving to the right with respect to O, and (which amounts to the same thing) O is moving to the left with respect to O'

The relation between the coordinates attributed to any given point by the observers in the two systems is:

$$x' = x - Vt$$

with $y' = y$ and $z' = z$. Of course, t is the time measured from the instant when the two observers were at the same place.

Since V is constant, the transformation is linear. For any phenomenon involving forces the term $-Vt$ in the equation disappears, since forces produce accelerations and accelerations are second order derivatives of the coordinates with respect to time (the second derivative of $-Vt$ is, of course, zero). In other words: the laws of Physics (or at least of mechanics) are formally the same for both observers, each one using his/her own coordinates.

Now, if you write down Maxwell's equations in their classical form in the coordinates of a given inertial observer, and then go to the reference frame of a different inertial observer, who is moving with constant velocity with respect to the former, you must perform a coordinate transformation. If the transformation is Galilean, the resulting equations will be different from the ones you started with, since additional terms due to the coordinate transformation will appear. Hence the laws of electromagnetism are not the same for all inertial observers, in contradiction with the principle of *Galilean Relativity*. Apparently, a preferential reference frame exists, in which the equations have their simplest form. An observer investigating electromagnetic phenomena should be able to detect her/his *absolute* motion with respect to it. In other words, all inertial observers are equal, but one *is more equal than the others*.

The idea of a special reference frame comes out of classical electromagnetism also for another reason. Maxwell's equations tell us that light is an electromagnetic wave, and waves are waves of *something*. However, light travels across empty space, such as between the sun and the earth. Therefore, as we have seen in Chap. 6, scientists at the time were led to believe that *something* (the *ether*) supporting the propagation of electromagnetic waves permeated the whole universe.[2]

In fact, the ether was a privileged reference frame, breaking Galilean relativity. The speed of light was intrinsically related to it, as the speed of sound is related to the air, or water, or any other medium in which it travels. In practice, the speed of light would have its typical $c = 299{,}792{,}458$ m/s value[3] with respect to this fundamental ether reference frame. Then the speed of light, measured by an observer moving with respect to the ether at a speed V, was expected to be different from c:

---

[2]Today we simply say that this *something* is the electromagnetic field (which is not a mere verbal expression), but in the 19th century people looked for some hypothetical substance, to which a name from ancient natural philosophy was given, i.e. *aether* (or, less classically, but in a simpler form, *ether*), or *fifth essence*, as the element filling the skies. The other four elements were: earth, water, air and fire.

[3]Today this is assumed to be a universal constant rather than a quantity to be measured. We have discussed the velocity of light in Chap. 3.

for a motion in the same direction as light, it should have been $v_{light} = c - V$; in the opposite direction $v_{light} = c + V$. This is what intuition and the Galilean transformations tell us.

If that were the case, light would be a perfect sensor to measure the absolute motion of an observer with respect to the ether. Of course, physical speeds V are usually much smaller than c, so that the expected change of the velocity of light for different observers would be extremely small. However, at the end of the 19th century, interferometric techniques with a high enough sensitivity to sense the absolute motion of the earth with respect to the interplanetary ether ($V_{earth} \sim 30$ km/s; $V_{earth}/c \sim 10^{-4}$) were already available. This was the basis of the famous Michelson and Morley experiment, or rather experiments, since various attempts were repeated by the authors between 1881 (Michelson alone) and 1887. (We have already described the nature of these experiments in Chap. 6.) Afterwards other investigators repeated the same type of measurement in the first three decades of the 20th century.

As we have seen, the Michelson-Morley experiments revealed no motion of the observer's laboratory with respect to the ether. This null result started a debate on the properties of the ether, and its possible total or partial drag by moving bodies, but the problem remained unsolved.

In 1889 the Irish physicist, George Fitzgerald, and later, more formally, in 1892, the Dutch physicist, Hendrik Lorentz, remarked that the zero result of the Michelson-Morley experiment could be explained by assuming that moving bodies are contracted in the direction of motion. The contraction depends on the velocity of the body with respect to the ether, but not on its nature. Working from this remark, various scientists, including Poincaré, found that the Fitzgerald-Lorentz contraction could be obtained in a straightforward manner if one transformed the equations in the transition from one inertial observer to another, not by means of Galilean transformations, but by using other linear transformations that are now named after Lorentz. The Lorentz transformations account for the null result of the Michelson-Morley experiment and leave Maxwell's equations unchanged.

In the case of the two sets of coordinates of Fig. 7.1, the Lorentz transformations are given by:

$$\begin{cases} x' = \gamma(x - Vt) \\ t' = \gamma\left(t - \dfrac{V}{c^2}x\right) \end{cases}$$

where the symbol γ, the Lorentz factor, is given by

$$\gamma = \frac{1}{\sqrt{1 - V^2/c^2}}.$$

The Lorentz transformations were indeed intriguing, since nobody at the time could understand what they could possibly mean. They exhibited a non-trivial dependence on the relative speed V, as required, but they also had unexpected consequences, such as that time flows differently for different inertial observers, which is completely counterintuitive.

7.2   A Clerk in the Patent Office of Bern

Such was the situation in 1905 when a young clerk of the patent office of Bern arrived on the scene. His name was Albert Einstein. He had graduated in physics five years before from the Swiss Federal Institute of Technology in Zurich.

In 1905 Einstein published six papers, three of which were of paramount importance for physics. The one that laid the foundation of the topic we are dealing with in this chapter was entitled: *On the Electrodynamics of Moving Bodies* [2]. The name of the subject it covered was coined by Einstein himself: *The Special Theory of Relativity.*

Einstein started from the violation of Galilean Relativity by the Maxwell equations and the failure of all attempts to detect the absolute motion of the earth with respect to the ether. Assuming that there was no natural conspiracy to frustrate the measurements, to get out of the *impasse* he proposed that the speed of light *had to be exactly the same for all inertial observers*, thus becoming a universal constant, and that no ether was needed.

This bold assumption had an immediate implication: the correct coordinate transformations between inertial observers are the Lorentz transformations, and this ensures the invariance of the Maxwell equations. However, if this is indeed the case, the laws of mechanics have to be revised to guarantee that they are the same for all inertial observers. The most bewildering consequence of the Lorentz transformations is, as we have already remarked, that not only the space coordinates of an event, but also the *time coordinate*, depend on the observer.

Let us now leave the historical trail and try to summarize the essentials of special relativity. Einstein removed the need for the ether but introduced a four-dimensional continuum as the arena in which physical phenomena are played out. That four-dimensional continuum is called *space-time*, and includes time on the same footing (almost) as any space coordinate. In space-time, mechanics is described by the geometric properties of the continuum. These properties differ from those of Euclidean geometry because of the constraint of invariance of the speed of light. The formal definition of this peculiar geometry was worked out by one of Einstein's teachers in Zurich, the Lithuanian mathematician, Hermann Minkowski, who gave his name to the space-time of special relativity.

We may illustrate the *new geometry* by drawing in Fig. 7.2 a bi-dimensional image of Minkowski's space-time for the case, analysed before *classically* (see Fig. 7.1), of the two observers O and O'. Only one space coordinate is shown, together with the time axis. In order to have homogeneous quantities,[4] the vertical

---

[4]Homogeneous quantities have the same dimensions and are measured in the same units.

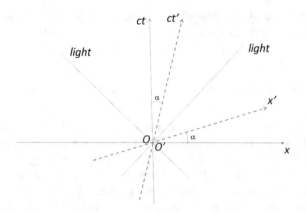

Fig. 7.2 Schematic representation of a bi-dimensional Minkowski space time mapped by an inertial observer O sitting at the origin of the {*x*, *ct*} coordinate system. The {*x*′, *ct*′} coordinates and axes belong to an inertial observer O′ in uniform motion with speed *V* in the positive *x* direction. At time *t* = *t*′ = *0*, the two observers coincide

axis does not represent *t*, but rather *ct* which is a length (remember that *c* is now a universal constant).

We define the *world-line* of a point as the path travelled by it through space-time as its history unfolds. Since the observer *O* sits, by definition, at the origin at all times, his/her *world-line* coincides with the vertical time axis. Light rays correspond to straight lines at 45° to the axes. In Fig. 7.2, two rays corresponding to forward and backward propagation along the x-axis are shown. When we consider all three space coordinates, the light rays passing through the origin describe a cone: the so-called *light cone*. This cone has two halves: one towards the past, the other towards the future.

In the same figure, the world-line of the second observer *O*′, who is in uniform motion at speed V with respect to O, is represented by a *broken* straight line at an angle α with the time axis. The angle α is given by:

$$\tan \alpha = \beta = \frac{V}{c}.$$

In the *proper* reference frame of *O*′, where he/she is at rest, the world-line of the observer *O*′ coincides with the time axis *ct*′. Now remember that the speed of light is the same for both observers, which means that the light cone must also be the same for both. Since the light rays bisect the angle between the time and the space axes, we see that the space axis of *O*′, when viewed from the viewpoint of *O* (in her/his reference frame), is a straight line symmetric to *ct*′ with respect to the light cone. This is the dashed line *x*′ of the figure.

In practice, the geometric approach shows that the Lorentz transformations resemble rotations, although not rigid rotations, such as the ones in space, but rather rotations of the axes towards the invariant light cone. The higher the relative speed

V, the larger the α angle, and the closer the axes of the moving observer appear to the light cone when viewed by the fixed observer. There is nothing special about the reference frame of the *moving* observer. Since motion is always relative, for observer O' the axes of the other observer O appear to close in towards the unique light cone by exactly the same angle α. Using a geometrical approach, the Lorentz transformations can be expressed in terms of rotations by imaginary angles.[5]

7.3   Paradoxes

The special theory of relativity has many consequences that are perfectly rational, though conflicting with our intuition. For example, let us consider the contraction of length implied by Lorentz transformations. From classical physics we learn that the contraction of a material rod requires an external compression force, whose effect is different for each material. The same force does not produce the same shortening on an iron rod as on a wooden rod or a rubber rod, but this is precisely what Einstein's Relativity predicts.

The geometric approach tells us that *no force* is needed to produce this effect, which is really a *projection*, rather than a contraction. The Lorentz 'contraction' is not the effect of some mysterious force, but depends on the choice of the space-time axes, i.e. on the observer(s) involved. The situation is readily clarified by a graph in Minkowski space-time (see Fig. 7.3), in which a rod is at rest in the reference system of the observer O' who, just as in Fig. 7.2, sits at the origin of the axis x'.

In ordinary three-dimensional space, the projections of a rod onto a pair of orthogonal axes have lengths, which depend on the angle between the rod and the axes. If the axes are rotated, the lengths of the projections change, even though the rod is not physically affected. The same happens in space-time. As we have seen, a Lorentz transformation between a pair of observers in relative motion is the analogue of a rotation. Each observer measures the *proper* length of a bar *only when* the bar is at rest with respect to the observer; otherwise what is measured is only a projection, which is always shorter than the proper length. The actual shortening is given by the inverse of the γ factor,[6] which corresponds to a rotation by an imaginary angle.

Clearly the same effect can be expected to happen with time intervals. Relying once more on graphics for a visualization of it, we turn to Fig. 7.4, which is similar to the previous two, but focuses on time measurements. *Points* in Minkowski space-time are called *events*. We have a *proper* time interval between a pair of events when both of them occur at a fixed *position* with respect to the observer. This is the case for events A' and B', which happen at the same position ($x' = 0$),

---

[5]An *imaginary* number is a multiple of i, which is defined as $\sqrt{(-1)}$ in the mathematics of complex numbers.

[6]$L = L_0/\gamma$.

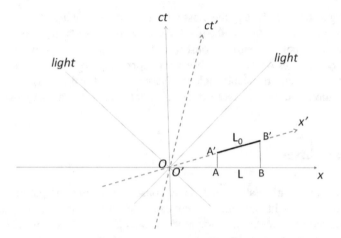

Fig. 7.3 Same as Fig. 7.2. The length L_0 of a rod is measured along the x' axis, i.e. in the reference system of the observer O' in which it is at rest. Therefore L_0 is the proper length of the rod. The projection of L_0 onto the x axis of observer O appears shorter

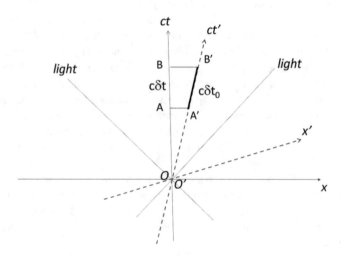

Fig. 7.4 The time interval between two events A' and B', at rest in the reference frame of O', appears shorter when projected onto the time axis of O

occupied by O' in his/her own reference frame. The corresponding time interval from the viewpoint of O is obtained by *projecting* the proper time interval between A' and B' onto the time axis ct used by O. For instance, if the time interval between A' and B' measured along the ct' axis is six seconds, the time interval between A and B measured along the ct axis is shorter, e.g. five seconds. The time interval for O is thus *shorter* than for O', although the physical phenomena taking place in A' and B' are not affected in any way.

Whatever happens in the moving frame is seen by O as taking longer than if it happened at rest in its own reference frame. This effect is known as *time dilation*. As for the amount of the dilation, it is given by the γ factor, just as was the case for the length contraction.[7]

Once more everything is clear from the viewpoint of geometry, but disconcerting from the viewpoint of our intuition.

Geometrically the relevant quantity associated with a pair of events is the space-time 'distance' between them. In ordinary space and using Cartesian coordinates, the (squared) distance between two arbitrarily near points is[8]

$$dl^2 = dx^2 + dy^2 + dz^2$$

In space-time the analogue squared 'distance', which is called the *interval*, embodies the invariance of the speed of light and is

$$ds^2 = c^2 dt^2 - dl^2. \tag{7.1}$$

The order of the signs in the above equation (which term is positive and which is negative) is conventional. In many books and scientific articles you may find the opposite choice (i.e. $ds^2 = -c^2 dt^2 + dl^2$). However, provided you preserve consistency with your choice when doing calculations, nothing changes from the physical point of view. On the formal side, if we adopt imaginary variables either for space (in the case of the choice adopted here) or for time (in the case of the opposite choice), the interval appears as an application of Pythagoras theorem in four dimensions. In fact, make the substitution

$$x \to i\xi; \quad y \to i\eta; \quad z \to i\zeta$$

and you will get

$$ds^2 = c^2 dt^2 + d\xi^2 + d\eta^2 + d\zeta^2$$

Considering (7.1), we see that if the two events are such that it is possible to travel from one to the other at a speed smaller than the speed of light c (i.e. $dl = V dt$, with $V < c$), then $ds^2 > 0$, and the interval is called *time-like*. If the two events are located along a light ray ($V = c$), then $ds^2 = 0$, and this is known as a *null* interval. If the required speed is greater than c, then $ds^2 < 0$ (despite the square sign) and the interval is *space-like*. However, as we shall see further on, it is impossible for any physical entity (be it matter or energy) to travel at a speed larger than the speed of light, which

---

[7]$T' = \gamma T$.

[8]Here the symbol of the differential operator d indicates the arbitrarily small difference between the values of a given variable between the end points of a segment. The third spatial dimension is included by adding an extra term, dz^2, to the equation.

Fig. 7.5 The proper time
elapsed from A to B for twin *1*
is *longer* than for twin *2*

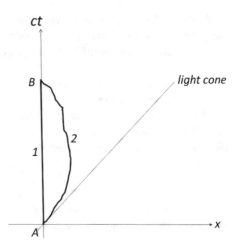

means that an event separated from a previous one by a time-like interval can be affected physically by the latter (something can indeed go from the first to the second) and they are *causally* connected. If instead the interval is space-like, no interaction can happen between the two events and they are *causally disconnected*. Incidentally, if we consider an inertial observer O' travelling at a continuously increasing speed V' with respect to O, we see that as V' approaches c, the Lorentz factor γ diverges (i.e. becomes infinite), thereby making it impossible to use the Lorentz transformations. This is a first hint of the impossibility of faster-than-light travel. A more physical and stringent argument will be presented later.

If one studies events happening at the same position in space ($dl = 0$) then $ds^2 = c^2 dt^2$, so we see that the interval is proportional to the proper time span, i.e. to the time read on the watch of the inertial observer at rest in the given reference frame.

The properties of Minkowski space-time, and in particular the time dilation, can explain one of the most famous '*paradoxes*' of special relativity: the so called *twin paradox*. One twin stays at home; the other travels through space at various speeds, then finally comes back home. When the twins meet again, the astronaut looks younger than his sedentary brother. The situation is once more explained graphically in Fig. 7.5.

The path followed by twin 1 to go from A to B is visibly shorter than that followed by twin 2. However the figure, being on a page, is necessarily Euclidean. The corresponding Minkowski situation must include the effect of that little minus sign contained in the definition of the interval, whose consequence is that world-line 2 from A to B is actually *shorter* than world-line 1. This means that the proper time elapsed along world-line 2 is shorter than the proper time span along world-line 1, so that the twin 2 in B is younger than twin 1. Everything is consistent though bewildering to our common sense. It is important to stress that in the example above the symmetry between the twins is broken. If they travelled at a constant speed with respect to each another, they would both be in inertial motion.

Fig. 7.6 Two events A and
B occur in the sequence A,
B for observer O, and B, A for
observer O'

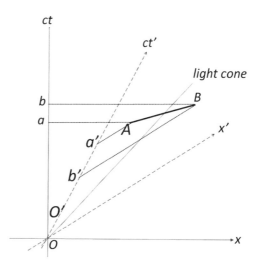

By exchanging light signals they could verify the occurrence of the time dilation:
each twin would feel that the other's time flows more slowly (everything takes more
time) than his/her own and there would be complete symmetry. However, if they
wished to compare their clocks in the same location, they would have to meet. One
of them (twin 2 in the above example) would have to undergo some acceleration,
changing both direction of motion and relative velocity. The periods of acceleration
would break the symmetry: for a while, twin 2 would cease to be an inertial
traveller, unlike twin 1. This asymmetry would introduce a physical difference
between them.

Another famous 'paradox' concerns simultaneity. Take a pair of events A and B,
separated by a space-like interval. For observer O, A happens *before* B. Now
examine Fig. 7.6, which includes a second observer O' in motion with respect to
O. A projection[9] of the same interval as before onto the axes of O' tells him/her that
B comes *before* and not after A. The inversion of the temporal order for different
observers is disconcerting, because of its implications on causality.[10] However, we
should remember that the interval between A and B in Fig. 7.6 has been selected to
be space-like, which means that no physical interaction between the two events can
exist, since it would need to be superluminal (travelling at a speed higher than c). In
other words A and B cannot be causally connected.

A special case of the above situation is the relativity of simultaneity, which
means that events which are simultaneous for one observer cannot be simultaneous
for another inertial observer moving with respect to the first.

---

[9]Remember that to project onto the time axes of O and O', we drop lines from A and B parallel to
the x and x' axes respectively.

[10]Causality is where the order of events is fixed by one event being 'caused', by the other. An
example is birth and death. One cannot die before being conceived.

From the moment of its publication, Einstein's relativity raised both enthusiasm and fierce opposition, even from within the scientific community. A famous example is the case of Georges Sagnac, a French physicist, who in 1913 discussed and measured an effect that still bears his name, related to closed paths of light (especially in rotating systems, but also in the case of inertial motion) [3]. He interpreted this effect as evidence of the existence of a *luminiferous* ether and as a disproof of the theory of relativity. Today the Sagnac effect is known to be a relativistic effect that has considerable practical importance in devices used to measure rotation rates, such as ring lasers, and in global positioning systems (GPS).

To conclude, in spite of all the clamour and folklore surrounding it, it is important to stress that the subject of the present chapter does not consist of amusing formal speculations *pour épater les bourgeois* (to shock the bourgeoisie). Instead it represents a vast and consistent body of physical phenomena, accurately and recurrently verified by experiments, ranging from the measurement of the lifetime of muons arising from secondary cosmic radiation, to the relativistic corrections in the GPS, which, if neglected, would lead to errors of more than a hundred kilometres in positioning.

7.4   The Most Famous Formula of Physics

In the same *annus mirabilis* 1905, Einstein published a second paper on the newly formulated theory of relativity, in which he proposed a formula, which is probably the most famous in the world, to the extent that it is now found everywhere, including on T-shirts. We are referring of course to the equation:

$$E = mc^2$$

where E stands for energy, m for mass. c, as we know, is the speed of light in a vacuum. Given the relevance of the formula, a formal *derivation* of it would be preferable, but we omit it here to avoid mathematical complexities. Instead we offer a simpleminded argument of *plausibility* and discuss its meaning, and above all its consequences. However, first we must make a short digression.

We have already repeatedly discussed the relationship between mathematics and science. Here we wish to stress a basic difference. Mathematics begins with a given set of assumptions, without regard to any physical reality, and derives consistently the consequences of these assumptions, usually by means of theorems. However, in spite of the usefulness of these theorems, the outcome is nothing new, in the sense that it is already implicitly included in the initial assumptions.

Science, on the other hand, begins from the available data, based on what we call *reality*. If the data do not suffice, we try to collect more by means of experiments. The goal of science is to arrive at some law, which not only must be consistent with the old data, but must also be capable of predicting some new phenomenology. However, we cannot really claim to demonstrate any such law (or theorem), since the concept of demonstration implies an absolute (universal) validity, while our range of experience is necessarily limited. Thus Newtonian Mechanics is superseded by Relativity, when velocities close to c are considered, or by Quantum Mechanics (as we will see in the next chapter) when one has to deal with very small objects.

The above discussion is, of course, naïve and simplistic, but it defines the boundaries of our expectations: a harvest of predictions, but no real demonstration, just some plausibility argument, which may be more or less sophisticated (or elegant). The former would be the formal derivation, mentioned above, the latter is what follows in its place.

We have already seen that the spatial and temporal distances between two events depend on the Lorentz factor γ, which in turn is a function of the ratio V/c. Likewise we can expect that the mass m of a particle moving with velocity V also depends on γ. Let us then explore the *Ansatz*[11]:

$$m = m_0 \, \gamma$$

where m_0 is the rest mass of the particle, i.e. its mass when the velocity is zero. Recalling the definition of γ and expressing it as a Taylor expansion (see Chap. 2)—which is well justified when V $\ll$ c, as is usually the case—we obtain:

$$m = m_0 \left(1 + 1/2 \, V^2/c^2 + \ldots \right).$$

Keeping only the first two terms of the expansion and multiplying by c^2, we obtain:

$$mc^2 = m_0 c^2 + 1/2 \, m_0 \, V^2.$$

The second term on the right is immediately recognizable as the traditional formula for the kinetic energy of the particle. From this identification follows the generalization that interprets the difference $m_0 c^2 (\gamma - 1)$ as the full relativistic kinetic energy. The quantity $m_0 c^2$ is also an energy, which is present even when the particle is not moving, and so must be the particle's *rest energy*. Finally, $mc^2 = \gamma m_0 c^2$ is interpreted as the *total* energy of the object. The quantity $m = \gamma m_0$ is called the *relativistic* mass of the body.

In practice, the inertial mass[12] of a moving body, when measured, turns out to be larger than its rest mass and *goes to infinity when V approaches the velocity of light*. This divergence implies the impossibility of accelerating any object up to the speed of light, since this would require an infinite amount of energy.

---

[11]An *Ansatz* in mathematics is an assumption about the form of an unknown function which is made to facilitate the solution of some mathematical problem.

[12]The inertial mass is the mass that appears in dynamical equations, in contrast to the gravitational mass that is the source of gravity. This will be clarified in the next section.

The customary boldness of the young Einstein led him to generalize his formula $E = mc^2$ further, including in E *any kind of energy*. The implications of this assertion are of paramount importance. Trivially we may remark that a sealed container of water weighs more when the water is hot than when the water is cold, because in the former case there is a contribution to the weight from the thermal energy of the liquid. The difference is so tiny that there is no way to measure it; however there are other situations where the contribution from pure energy is much more relevant.

If we consider any physical system in which attractive interactions are at work, we know that the corresponding interaction energy is *negative,* and so also is the inertial (binding) mass that the interactions contribute to the total mass of the system. An atom has a smaller mass than the sum of the rest masses of the nucleus plus the electrons that form it; a nucleus has a total rest mass smaller than the sum of the rest masses of its component nucleons. The latter fact opened the way to the most powerful and terrifying application of $E = mc^2$ and gave an important contribution to the fame of Einstein. The importance of $E = mc^2$ for nuclear physics will be stressed also in Chap. 9.

On one side of the Mendeleev periodic table of the elements we have heavy nuclei which tend to be weakly bound and unstable; examples are uranium isotopes such as U^{238} and especially U^{235}. When one such nucleus undergoes fission (i.e. splits into parts), the fragments are smaller and more tightly bound nuclei, so that the total final mass of the fragments[13] is smaller than that of the initial nucleus. The excess mass shows up in the form of kinetic energy, which is mostly manifested as thermal energy. This elementary phenomenon is the basis both of nuclear power plants (where fission takes place under controlled conditions) and of atomic fission bombs.[14]

At the other end of Mendeleev's table we find the lightest elements (such as hydrogen and helium). There, if one succeeds in pushing two nuclei of the same species one against the other, overcoming the electrostatic repulsion, they 'fuse', i.e. attach to each other because of the dominating nuclear interaction. The new compound nucleus turns out to be more tightly bound than the original two. Once more the excess mass is converted into energy, both in the form of photons (and neutrinos too) and of kinetic energy of the final nucleus. This process is even more efficient than nuclear fission. It is the basis both of the hydrogen bombs which are still stocked in tens of thousands in the military arsenals of the main super powers, and of the energy continuously released in the stars, and in particular in our sun. (We will deal with nuclear fission and fusion in more detail in Chap. 9.)

As we see, relativity is certainly not just a simple fanciful theory for a handful of hare-brained scientists.

---

[13]Including a few free neutrons and photons.

[14]Such as the ones dropped on Hiroshima and Nagasaki at the end of the Second World War.

7.5 General Relativity

It took ten years for Einstein to progress from his theory of special relativity to the *theory of general relativity* (GR). In 1905, the only known fundamental forces were electromagnetism and gravity. Electromagnetism was splendidly accounted for by the special theory. However, this was not the case for gravity.

Newton's theory of gravitational interaction had been an outstanding achievement and was able to explain perfectly well the celestial mechanics within the solar system, as well as the dynamics of falling bodies on earth. Nobody really bothered at the time about a few *marginal details,* such as the anomalous precession rate of the perihelion of Mercury.[15] It was expected to be explained, sooner or later, by the slightly non-spherical shape of the sun and/or the gravitational perturbation due to other bodies in the solar system.[16] Yet, something was unsatisfactory, at least from a conceptual point of view.

The formulae of Newton's universal gravitation resemble those of electrostatics, but in the case of the electromagnetic interaction, the full electromagnetic field is required. With gravity the duality between electric and magnetic components was apparently missing. The electromagnetic waves, e.g. light, were known to travel at an extremely high, but *finite* speed. In Newtonian gravity no analogue of the magnetic field exists and it is not possible to write any wave equation, as is the case in Maxwell's theory. A consequence was that any change in the distance appearing in Newton's gravity formula, i.e. the movement of any massive body anywhere in the universe, would produce an *immediate* effect on earth. This amounts to saying that the gravitational interaction apparently should propagate at infinite speed and an absolute simultaneity would be established throughout the whole cosmos.

On one hand, the idea of an infinite propagation speed for gravity may sound somewhat unphysical, but on the other its implied absolute simultaneity is consistent with Newton's concept of absolute time. It contradicts, however, the special theory of relativity and, in fact, any attempt to treat gravity as a field (similar to electromagnetism) in a Minkowskian background failed.

At the beginning Einstein needed some clue from which to start his task, and lacked some important mathematical tools that he acquired afterwards, helped by his friend Marcel Grossmann. He started from the old remark, already made by Galileo, that the inertial mass coincides with the gravitational mass. This assertion is known as the *equivalence principle.* In other words the parameter that accounts for the 'matter content' of a body coincides with the parameter measuring the strength of the gravitational attraction. In electromagnetism, on one side of the

---

[15]The orbit of Mercury about the sun is an ellipse (approximately). The point of the ellipse's closest approach to the sun is found to rotate (or precess) slowly around the sun.

[16]Le Verrier in 1859 had proposed the presence, inside the orbit of Mercury, of another unseen planet christened *Vulcan.* Others had suggested the existence of an equally unseen satellite of Mercury.

equation we find the inertial mass of an object, on the other its electric charge. Why is the *gravitational "charge"* equal to the inertial mass?

A consequence of the above coincidence is that the free fall of an object in a gravitational field does not depend on the mass of the object; a cannon ball and a feather fall with exactly the same acceleration in a vacuum. This fact suggested to Einstein a thought experiment. Suppose you are in an elevator. Suddenly the rope suspending the elevator breaks, and everything starts falling freely with the same acceleration. If you had been asleep and awoke at that moment, you would find yourself floating in the air, with all your chattels drifting about you. You would not be able to say whether there is a gravitational field, or not. If you were a physicist and carried out some experiment, you would obtain the same results as if there were no gravitational field at all.

Now consider a situation complementary to the one we have just described. You are sleeping on a space ship freely floating far away from any gravitational field. Everything is in inertial motion. Somebody then fires the engine, which imparts a uniform acceleration to the ship. When you awake, you will find yourself pressed against the floor of the cabin (of course if it is on the side of the engine) and everything you drop will fall to the floor with the same acceleration. How can you know whether this is due to the regular push of an engine, or to a real gravitational field?

With his usual perspicacity, Einstein concluded that:

It is impossible to distinguish, *locally*, a gravitational field from a uniform acceleration field;

A freely falling observer is *locally* indistinguishable from an inertial observer, so that the same laws of physics apply to both of them.

The adverb *locally* used in both statements means that the conclusions are true provided the volume you consider around you is not too large, in the same sense that you are unable to say that the surface of the earth is curved when you examine a limited portion of it (in practice, when the transverse size of the area is small with respect to the radius of the planet). Both above sentences express the *equivalence principle* for gravitational interaction.

Recall now that special relativity reduced the description of physics for inertial observers to geometry in four dimensions on a flat manifold. The term "manifold" has here a technical meaning typical of geometry; we may understand what it is by a familiar analogy. In our ordinary world, and in two dimensions, a manifold is a surface, on which you may draw figures and study their geometric properties. A surface can be flat, like the top of a table, or curved like the boundary of a sphere. In the former case, the geometry on it is Euclidean; in the latter, the geometry is Riemannian.[17] Adding dimensions and preserving the same logic we have higher dimensional manifolds that are the analogue of ordinary surfaces. The space-time of special relativity is a four-dimensional *flat* manifold, on which the geometry is Minkowskian.

---

[17]From the name of the German mathematician Bernhard Riemann.

To make a long story short, let us jump straight to the conclusion: space-time is still the background for any physics, but it is not necessarily flat. When the gravitational interaction is included, the four-dimensional manifold turns out to be *curved*; we shall further comment on the curvature in a moment.

What we have learned above about inertial motion in a flat manifold can be generalized to motion on a curved manifold. In flat space-time, an inertial observer moves along a straight world-line. His/her analogue falling freely in a gravitational field will move along the analogue of straight lines in curved manifolds. These are known as *geodetic* lines. Leaving all formalisms and subtleties aside, we may say that a geodetic is the shortest path between two points on a curved surface (in our case, the four-dimensional manifold).

The shape of the geodetics depends only on the curvature and shape of the manifold upon which they are drawn, not on the nature and mass of the falling object. Furthermore, excluding for the moment the presence of singularities, cusps and the like, any given surface is locally indistinguishable from a plane tangent to it. In our case, the curved space-time (the curved four-dimensional manifold) is locally indistinguishable from its *tangent space*, the flat Minkowski space-time. This is the geometric motivation of the equivalence principle.

The idea of a curvature of the four-dimensional space-time may be rather confusing. Does it require the existence of a fifth dimension from which to look at it? In fact, in our every-day experience, we are used to looking at the curvature of an ordinary 2D[18] surface from the third dimension in space. Also, what does it really mean that a 3D space (or manifold) is curved? Or, worse, a 4D space?

Geometry, however, does not necessarily require additional dimensions to establish the presence or absence of curvature. Suppose you are flat on a surface and able to draw triangles on it. When you measure the internal angles of the triangles, if you find that their sum is π (as in Euclid's geometry), you conclude that the surface is flat. However you could also find that the result is $>\pi$, as would be the case upon the surface of a sphere (or less than π, as would happen on a saddle shaped surface). So we see that, in order to detect the curvature of a manifold, there is no real need for an additional dimension.

What about the laws of physics in the presence of a gravitational field? For a freely falling observer, *locally* the laws look identical to those for an inertial observer. In practice, given any particular choice of coordinate system, the laws are usually expressed in the form of partial differential equations (see Chap. 2), describing the behaviour of tensorial quantities.[19] The general form of these equations must be such that they reduce to the above mentioned partial differential equations, when they are restricted to a small enough region (technically a four dimensional volume) close to the origin of the reference system, where the freely falling observer sits. In practice, one must combine the changes involving the

---

[18]nD means n-dimensional.

[19]Tensors are arrays of numbers corresponding to physical quantities, endowed with peculiar transformation properties when changing the reference frame. A vector is a rank-1 tensor.

relevant physical quantities in a given inertial frame with the fact that the frame itself is also rotated and stretched when moving around on a curved manifold.

Once we have learnt how to express the physical laws in the presence of gravity, i.e. in a curved space-time, there remains one further basic question: what produces the curvature of the manifold? Unsurprisingly, the source of the curvature is the source of the gravitational field in Newtonian physics, i.e. mass, or more precisely *mass/energy*, now that we have established the equivalence of the two. An imaginative and concise way to represent the gravitational interaction can be found in a famous quotation from John Wheeler: "*Space-time tells matter how to move; matter tells space-time how to curve*". Mathematically the link between space-time and mass/energy is contained in Einstein's equations, which are a set of ten second-order partial differential equations.

The solution of the Einstein equations is in general not easy. The number of closed formal solutions is rather small and these deal with situations with strong symmetries, which help to reduce the number of independent equations. It is beyond the scope of the present book to enter into the details and discuss the different solutions. In the following we will limit ourselves to mentioning two relevant examples: the Friedmann-Le Maître cosmological solution and the Schwarzschild static solution.

Starting with the former, let us consider a simplified description of the visible universe, treated as a homogeneous and isotropic *dust*, made of galaxies,[20] and then try to *solve* Einstein's equations. The big surprise, which Einstein was initially unwilling to accept, was that the only stable solutions implied that the universe must be either expanding or contracting. Astronomical observations at the end of the twenties confirmed that the universe is indeed expanding.

As for the Schwarzschild solution, it holds when the source of curvature is spherical and static. The weak field limit of the solution is Newton's gravity for a spherical mass. Setting the details aside, a most interesting feature of this solution is that, when the mass density of the source exceeds a given limit,[21] a very peculiar object, known as a *black hole*, appears. The black hole is delimited by a horizon from which not even light can escape. Under the horizon, matter collapses towards a *singularity*, i.e. an infinitesimally small volume with an infinite density. Infinities and singularities signal trouble in various areas of physics, but in reality they are out of its domain, since in the real world (as distinct from the mathematical world) scientists do not know what to do with infinities. However, perhaps as a stroke of luck, the presence of the horizon prevents us from seeing what really happens inside a black hole. Facetiously, this fact has been called the 'cosmic censorship' principle.

---

[20]Actually in the first couple of decades of the 20th century people, including Einstein, thought rather of a dust of stars. The idea of a galaxy as an 'island universe', though lingering since the 18th century, was not considered.

[21]To have an idea of how high the density must be, consider that it should be higher than that corresponding to squeezing the mass of the whole earth into a volume of radius not more than 8 mm.

7.6   Sounds from the Depths: Gravitational Waves

While this book was being finalised, on the 11th of February 2016, the
LIGO-VIRGO[22] collaboration officially announced what had already been circu-
lating in the form of rumours since September 2015: for the first time a transient
gravitational wave (GW) had been detected by the two LIGO interferometers
located respectively in Hanford (Washington, USA) and in Livingston (Louisiana,
USA) [4]. This is an important discovery and is confirmation of Einstein's GR.

In order to explain what a GW is, let us go back for a while to waves in general.
Physically, a wave is a field depending both on space and time in such a way that
energy is carried from a source, where the wave is generated, towards infinity, or in
this case, vice versa (i.e., instead of a source we consider a final receiver which
absorbs the energy carried by the wave).

We have mentioned waves many times in this book. The more intuitive and
familiar kind are the elastic ones (typically sound waves). They travel through
elastic media, where the force attracting each elementary component to its neigh-
bour is proportional to the distance.

In the case of electromagnetism, the electric force between a positive and a
negative charge is inversely proportional to the square of the distance, and if that
were the end of the story, no wave (i.e. no net transfer of energy through empty
space) would exist. However, as we have seen in Chap. 6, whenever we have
moving electric charges, we also have a magnetic field, and Maxwell's equations
show that the full electro-magnetic field is the solution of a typical wave equation.
Waves are produced when the source (charge) is, for some reason, undergoing an
accelerated motion; if so, energy is radiated away and subtracted from the motion
of the source. A simple example is a radiating dipole, which may be visualised as a
pair of oppositely charged particles held apart by a "spring". If the spring starts
oscillating, the two moving charges radiate an electromagnetic wave, thereby
subtracting energy from the oscillation of the spring until the dipole (asymptotically) returns to the static equilibrium position (unless some external energy source
keeps on feeding mechanical energy into the system).

Now what about gravity? If we consider only Newtonian gravity (i.e. a $1/r^2$
attraction) no waves are possible.[23] The situation is however different in GR. As we
know, in this case the gravitational field in vacuo satisfies the geometro-dynamic
Einstein's equations mentioned in Sect. 7.5. The relevant quantity in these equations is the metric tensor (a four × four matrix).

We have mentioned tensors briefly already. Their exact nature—they are a type
of higher order vector—is not important here. If no gravitational source (i.e. mass)

---

[22]LIGO stands for Laser Interferometer Gravitational-wave Observatory. Virgo comes from the
cluster of galaxies in the Virgo constellation where scientists deemed it more probable to spot
some source of gravitational waves than from other parts of the sky.

[23]Not all time-dependent fields are waves. The peculiarity of a wave is to propagate at some typical
proper velocity, carrying energy only in the propagation direction.

is present, then space-time is flat and, using dimensionless quantities, the non-zero elements of the metric tensor are just 1's (corresponding to a Minkowski geometry in Cartesian coordinates). Even with a gravitational source present, because of the extremely weak nature of the gravitational field, the deviations of these tensor elements from 1 is very small. For instance, when we add the gravitational fields typical of the solar system, we see that the deviation from those 1's at the surface of the earth is in the order of 10^{-9}, and at the surface of the sun it is 10^{-6}. Of course if we go farther away those numbers decrease even further.

This smallness of the deviations from flatness allows us to simplify the full Einstein equations by treating the deviations as perturbations in a power series expansion of the relevant quantities, neglecting all higher order terms. This procedure is analogous to keeping only the first order terms in a Taylor expansion, as we discussed in Sect. 2.6. The interesting feature of the final result of this *linearization* of the equations is that they, in the end, look like the Maxwell equations for electromagnetism. In practice, things behave as if there were *two* components of the gravitational field: one similar to the good old Newtonian field, nicknamed *gravito-electric* field; an additional one similar to the magnetic field of the electromagnetic theory, dubbed *gravito-magnetic* field. The latter is in general very much smaller than the former.

In any case, the approximate correspondence of the linearized Einstein equations to the Maxwell equations also implies that gravitational waves should exist, travelling through empty space at the speed of light c and carrying gravitational energy away from some source towards infinity. Einstein was aware of this possibility as early as 1916. However he also declared that GW would probably never be detected because of their extreme smallness.

In considering possible sources of GW, the situation is a bit different from electromagnetism. The existence of positive and negative electric charges, implying attraction and repulsion, means that the simplest emitter is a time-dependent *dipole*.[24] In the case of GW, masses have only one sign and the interaction is always attractive; the consequence is that the simplest emitter is a *quadrupole*, rather than a dipole. A pair of similar massive objects orbiting around the common barycentre (i.e. the centre of gravity) is a time-dependent mass quadrupole, and is expected to radiate GW.

A positive confirmation of this part of GR theory was obtained for the first time after the discovery in 1974 of the first *pulsar* (a pulsating star, which does not actually pulsate, but rather emits a continuous beam along a direction that rotates about a fixed direction in space, in analogy to the light beam emitted from a

---

[24]A distribution of matter may be approximated as a series of elementary distributions. The simplest is the *monopole* distribution, where everything is concentrated at a point; then comes the *dipole*, where there are two points at a given distance from each other; next we have a *quadrupole*, corresponding to four masses at the corners of a square; then the *octupole*, for which the distribution is three-dimensional, with eight masses; and so on. The situation is analogous to a power series, with each term being smaller than the previous one and the whole series converging to the actual physical distribution.

Fig. 7.7 The decay of the orbital period of the binary system containing the Hulse-Taylor pulsar. The continuous line is the expected trend predicted from GR according to the quadrupole radiation of the system. The diamonds are the measured values (The origin of the figure is "By Inductive load—Own work, Public Domain," https://commons. wikimedia.org/w/index.php? curid=9538634.)

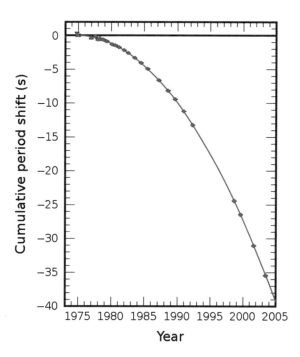

lighthouse). It is known as the Hulse-Taylor pulsar, from the names of its discoverers, but its technical name is PSR B1913+16, and it is located approximately 21,000 light years from the Earth. A pulsar is a compact object with a mass of the order of a few solar masses compressed into a sphere of radius in the order of 10 (or a few tens of) km. Astronomers have observed that PSR B1913+16 is part of a binary system and the companion, though not a pulsar, is also a very compact object, i.e. a *neutron star*.[25] This binary system is indeed a time-dependent quadrupole and it is possible to calculate the continuous GW emission rate from it. A consequence of the gravitational radiation should be a progressive decrease of the distance of separation of the two stars and a corresponding decay of the orbital period.

The binary system containing PSR B1913+16 continues to be monitored and observed since the time of its discovery. Not only has the orbital period decay been measured, but its time evolution is in fantastic correspondence with the behaviour predicted by the quadrupole emission formula of GR. Figure 7.7 shows the measured data and the theoretical prediction until 2005. The agreement is within 0.2 %.

Figure 7.7 is clear evidence of the emission of GW, but there is no hope of capturing those waves on earth because of the exceedingly small intensity of the phenomenon. If we want to detect the arrival of GW on earth we must look to other

---

[25]Objects of this type, as well as their active versions, i.e. pulsars, are the core remnants of massive stars after their explosion as supernovae.

stronger sources. In fact, viable candidates are restricted to catastrophic events where the energies at stake are much greater. The typical candidate event is a supernova explosion. The last two supernova explosions in our own galaxy were seen four centuries ago and took place in a time span of less than forty years. One was the 1572 event observed in particular by the Danish astronomer Tycho Brahé; the other was observed by Johannes Kepler in 1604.

Since these two events, nothing similar has occurred within the Milky Way. The closest external supernova took place in 1987 in the Large Magellanic Cloud, a small irregular galaxy of the so called "local group", of which the Milky Way is a member. Every day new supernovae are seen in other galaxies by astronomers but the distances are such that the expected GW intensity at the Earth is well below the sensitivity threshold of the GW antennas we have built so far.

To detect GW on Earth, one therefore has to look for even more powerful events, and these could be the merger of extremely massive objects. The most viable candidate involves esoteric objects such as black holes,[26] or to be more specific, pairs of black holes orbiting around a common centre of gravity and evolving towards the final merger, just as is the situation in the Hulse-Taylor binary system. The energy density in the former case, however, is so high, especially in the final stages of the merger, that the energy carried to our detectors is large enough to enable the measurement. The difficulty is of course that the frequency of occurrence of binaries composed of two black holes is incredibly small and, until a few months ago, one could even have doubted their very existence.

While waiting for an appropriate event to happen, theorists prepared "templates" reproducing the expected shape (the time dependence of the amplitude of the wave) in the various cases in order to recognize the signal when it arrived.

The signal recorded on September 14, 2015, and mentioned at the beginning of this section, corresponds to a merger of two black holes which happened approximately a billion years ago.

The antennas which have detected the event are large interferometers, similar in principle to the interferometer shown in Fig. 6.3, but with the two mutually perpendicular arms each 4 km in length. The passing GW modulates gravity slightly producing different effects on the two arms: one shrinks a little, the other extends a little. The difference is revealed by a change in the interference pattern of the returning light beams. To gain some idea of the sensitivity of this type of interferometer consider that the transient length difference between the two arms is of the order of 1 part in 10^{21}.

Incidentally, this major achievement of fundamental experimental physics is not only a confirmation of the existence of GW (already clearly indicated by the Hulse-Taylor pulsar binary) but is also evidence for the existence of binary black holes. GR continues its series of successes with the attainment of this new milestone in the progress of physics.

---

[26]Black Holes are discussed further in Sect. 11.7.

In the next chapters we will encounter numerous applications of Einstein's two theories (special and general relativity). For instance, modern nuclear accelerators incorporate relativity into their basic design, just as Newtonian physics is basic in the design of a motor car. Undoubtedly the two Einstein theories represent one of the pinnacles of intellectual achievement of the 20th Century. As one might expect, theories which are so counter-intuitive require a bold re-examination of several long-held philosophical beliefs. We will touch upon some of the philosophical implications of relativity in the final chapter of this book (Chap. 13).

References

1. G. Leopardi, in La Ginestra, Canti (1845)
2. A. Einstein, Zur Elektrodynamik bewegter Körper, Ann. Phys. 17(4), 891–921 (1905)
3. A. Tartaglia, M.L. Ruggiero, Am. J. Phys. **83**(5), 427 (2015)
4. B.P. Abbot et al., Phys. Rev. Lett. **116**, 061102 (2016)

Chapter 8
Quantum Mechanics

God does not play dice with the universe.
Albert Einstein, The Born-Einstein Letters 1916–55

Not only does God play dice but... he sometimes throws them where they cannot be seen.
Stephen Hawking, The Nature of Space and Time (1996)
by Stephen Hawking and Roger Penrose, p. 121

Abstract Quantum mechanics was developed at the beginning of the 20th century almost in parallel with relativity. It requires a high degree of abstraction and, usually, sophisticated mathematical tools. Reducing all formalism as much as possible, the accent is placed both on the counterintuitive and the novel concepts implied by the theory. The chapter starts again with Einstein's intuition applied to the explanation of the photoelectric effect. Then the quantum behaviour of sub-atomic particles is introduced. Further generalizing the analysis, quantum fields are touched upon and the open problem of the conflict between general relativity and quantum mechanics is considered.

8.1 A Disconcerting New Physics

Quantum Mechanics (QM) represents, perhaps even more than Relativity, a giant monument to human ingenuity and to the achievements of Science. At the same time, it is an unprecedented challenge to our capacity to understand, or even to define what *understanding* means. Richard Feynman, Nobel laureate for his fundamental contributions to quantum electrodynamics, once remarked: "*I think I can safely say that nobody understands quantum mechanics*" [1], which indicates perhaps the depth of the theory.

To approach QM we must keep our mind totally open and erase many of the thought patterns accrued through our everyday life and even through our scientific training. Hence it should not be surprising that one of the major contributors to the development of QM, Louis de Broglie, made his ground-breaking work (which we shall discuss in the following) during the course of his Ph.D. dissertation, when his head was uncluttered with the prejudices of the past, thereby obtaining the Nobel prize at only 37 years of age.

© Springer International Publishing Switzerland 2016
R. Barrett et al., *Physics: The Ultimate Adventure*, Undergraduate Lecture Notes in Physics, DOI 10.1007/978-3-319-31691-8_8

To describe our required state of mind when we approach QM, we might recall the verses of the greatest Italian poet, Dante Alighieri:

"Qui si convien lasciare ogni sospetto
Ogni viltà convien che qui sia morta."
("Here you must renounce your slightest doubt
And kill your every weakness.") [2]

Just as Dante had the Latin poet Virgil as a guide in his momentous journey through Hell and Purgatory, we also require a guide to visit *Quantumland*.[1] Unfortunately, the only adequate guide is mathematics, since it allows us to approach the most abstract concepts without a need to formulate them in terms of our limited everyday experience. To avoid mathematical intricacies, we shall have to present here some conclusions, even counterintuitive ones, without offering a rigorous explanation of them, and for this we apologize in advance. The reward for our undertaking will be a rich harvest of very elegant results, which defy the inventiveness of any science fiction writer, and yet are corroborated to an unsurpassed level by experimental evidence.

8.2   Quantization of Light

Leaving the epistemological issues aside for a while, the start of quantum mechanics, as well as of relativity, can be traced back to the end of the 19th century. As we have seen in Chap. 6, an unsolved problem at that time was the black-body radiation spectrum, and the discrepancy between the predictions of the classical theory of electromagnetism and the then current experimental observations. Such a discrepancy is for a physicist what a paradox is for a mathematician, i.e. an alarm bell that signals that *something is wrong* with the current theory, and therefore heralds an opportunity for a discovery.

As we have seen, a solution was found by means of Planck's more or less arbitrary assumption that the emission of radiation by any single oscillator (a molecule) takes place in finite packets of energy proportional to the oscillation frequency (*not* to the square of the oscillation amplitude, as would be expected in Maxwell's theory). To recap, the amount E of energy emitted by a molecule (henceforth called a *quantum*) must be proportional to the frequency ν:
   i.e.,

$$E = h\nu$$

---

[1]Robert Gilmore in his beautiful (and highly recommended) book "*Alice in Quantumland*" makes an allegory of QM as an "*extended analogy*" between the realm of QM and the nonsensical world encountered by the Alice of Lewis Carroll.

where h is a universal constant, now known as Planck's constant, whose numerical value is 6.626×10^{-34} m^{2}kg/s. With h being so small, *quantum effects* can be expected to play a significant role only at the molecular level or below, but the indirect consequences can also be extremely important macroscopically, as we shall see.

So the problem was formally solved, but the question remained: why should the energy be emitted only in discrete amounts? Planck conjectured that not only the emission, but also the absorption had to be by means of *quanta*, but again without any physical explanation.

As we have seen in Chap. 6, Planck's conjecture was confirmed, thanks to the intuition of Albert Einstein about a different phenomenon: the *photoelectric effect*. After his 1905 paper on this effect, for which he was awarded the Nobel Prize in 1921,[2] it became clear that electromagnetic radiation was both emitted and absorbed in finite and discrete quanta (later called *photons*). The Compton Effect, also mentioned in Chap. 6, provided further confirmation of Einstein's novel idea.

Thus, after approximately one century, the old Newtonian[3] interpretation of light as a flux of particles was re-emerging. However, the interference and diffraction phenomena, that had led scientists, ever since the time of Young's experiments, to describe light as a wave, were still there undisputed, leading scientists to the conclusion that in some phenomena light behaves as a wave, while in others it appears as a stream of relativistic particles all moving at the same speed c. Incidentally, from relativity the fact that photons travel at the speed c implies that their rest mass must be zero. No observer can run with the same speed as a photon, so nobody can examine a photon at rest!

8.3 Quantization of Matter

This 'split personality' of light has extremely important and totally unexpected consequences also for matter. In fact, the quantization of the electromagnetic energy emitted by an atom implies that its internal energy must also be *quantized*! Hence its internal energy cannot assume *any* value whatsoever, but only well-defined values, which are distributed along a ladder of unequally spaced steps according to the frequency of the photons that the atom can emit (or absorb). This was already well known to spectroscopists at the end of the 19th century. The light emitted by a pure element turns out to be a mix of a limited number of well-defined frequencies. Each chemical element has its own particular *spectrum of frequencies* in the light emitted by its atoms, to the point that the spectrum can be considered as the element's characteristic signature. But what does this mean?

---

[2]Einstein's Nobel Prize was not related to his work on Relativity, contrary to popular misconception.

[3]Strictly speaking, the idea of light as a flux of particles is much more ancient than Newton's time, going back to antiquity.

As we will see in Chap. 9, at the end of the first decade of the 20th century, scientists began to conclude that atoms are composite systems made up of a small massive and positively charged nucleus surrounded by a number of even smaller negative electrons revolving around the nucleus, in the same manner that planets revolve around the sun. It followed that the discrete distribution of energies in the spectrum hinted at an equally discretised distribution of radii (or shapes) of the electronic orbits. In other words, only a limited number of orbits are allowed and they correspond to well-defined orbital angular momenta (see Chap. 2).

Let us skip the long debate that engaged the physicists of that time concerning the quantization of the angular momentum, and remark that even in the macroscopic world, with which we have to deal every day, we observe various examples of '*quantization*'. For instance, if we pluck a guitar string, it will begin to oscillate and the half-wavelength of the oscillation will always be a sub-multiple of the total length of the string.[4] As a consequence only a discrete set of allowed frequencies[5] are excited. Similarly, if we strike a bucket containing water, we observe standing waves appearing on the water surface, i.e. in general, when we physically constrain a wave, its parameters, such as wavelength and frequency, become '*quantized*'. We will discuss further examples of this quantization in Chap. 9.

Let us now consider another phenomenon that was discovered experimentally in 1927. When a stream of mono-energetic electrons is shot through a very narrow slit,[6] the electrons scatter beyond the obstacle. The pattern of hits on an orthogonal plane located beyond the slit is not, as one might expect, distributed with a number of hits decreasing at larger distances from the centre of the projection of the slit, but appears to be made up of fringes, as you would expect from the diffraction of a wave. This behaviour had been predicted by the French physicist Louis de Broglie in 1924. Studying the double nature of the photon, de Broglie conjectured that also the electron (in fact any particle) has a double nature, i.e. displays both particle-like and wave-like properties. In analogy with the photon, the wavelength 'associated' with the electron must be inversely proportional to its momentum:

$$\lambda_e = \frac{h}{mV}$$

where V is the velocity of the electron with respect to the chosen (inertial) reference frame. At a low speed (with respect to c), i.e. in a non-relativistic approximation, m is the *rest mass* of the electron; otherwise it is the *dynamical mass* $m = \gamma\, m_{rest}$, where γ is the *Lorentz factor* introduced in the previous Chapter.

---

[4]We refer to 'long-lasting' oscillations. They are called the harmonics of the oscillating string.

[5]The frequencies, in first approximation, are inversely proportional to the wavelengths and depend on the nature and the tension of the string.

[6]In the real experiments made by George P. Thomson and, independently, by Davison and Germer, in 1927, a crystalline lattice was used. The diffracting openings were the regular gaps between the atoms of the lattice.

Pursuing the analogy, we find that the frequency of the electron wave is

$$v = \frac{E}{h} = \frac{mc^2}{h}.$$

As already mentioned, these considerations are not limited to electrons, but hold for any particle. Since the wavelength is inversely proportional to the mass and nucleons[7] have a rest mass almost 2000 times larger than electrons, their *associated wavelength* is 2000 times smaller and the corresponding diffraction phenomena are much more difficult to observe.

In conclusion, according to de Broglie's conjecture, matter, as well as light (photons), exhibits both a particle and a wave nature. The difference between the two is that the rest mass of particles is not zero. Particles have other parameters in common with photons, e.g. an *intrinsic* angular momentum, which is called *spin*, in analogy with the spin of a top.[8] After a detailed mathematical analysis, which we omit here, only two values (*eigenvalues*)[9] are found to be possible for the spins of electrons and nucleons. These values are: $\pm\hbar/2$. In other words, both electrons and nucleons have spin 1/2 (in units of $\hbar = h/2\pi$).

8.4   Wave Functions

The correspondence between particles (of matter) and photons, proposed by de Broglie, requires that a *wave function* be associated with each of them, to specify their *wave state* in addition to their *particle state*. In the case of photons, the waves have a well-defined physical nature as electromagnetic waves, whose explicit form can be obtained by solving Maxwell's equations. Their amplitudes are thus expressed in terms of electric and magnetic fields, which we can measure. But for particles, what does the *wave function* mean and how can we measure or calculate it?

Surprisingly it is easier to answer the third question first. A wave is in general expressed by some *functional form*, i.e. not a number, but a function ψ (as the wave function is usually called) of some *argument*. The wave function contains the information on the physical quantities carried by the wave. In the case of electromagnetic waves, the argument contains a frequency, proportional to the energy of the associated photons, and a wavelength, inversely proportional to their momentum.

---

[7]This is the name jointly given to the particles in the nucleus of an atom, i.e. protons and neutrons. See Chap. 9.

[8]The analogy should not be taken too literally. A top is an extended object and the proper rotation happens in three-dimensional space; the spin of an electron is an intrinsic parameter which goes to zero when we let $h \to 0$, i.e. when the quantum behaviour is negligible.

[9]Eigen comes from the German word meaning "own". An eigenvalue is the discrete quantized value of some property (in this case the spin angular momentum) of a quantum mechanical system.

We expect the same thing to happen for particles; however, since the propagation speed is a universal constant for photons but not for massive particles, an additional piece of information is needed and obtained from the conservation of energy, which we assume to be valid also in *Quantumland*. The conservation of energy requirement is introduced via the Hamiltonian, which we have already encountered in Chap. 4. Omitting the mathematical details, this procedure leads us to the *Schrödinger equation*, which is a second order *partial differential equation* (see Chap. 2), whose solution yields the functional form of the wave function.

Returning to the first two questions, i.e. the meaning of ψ and how to measure it, the answer is disappointing (even for scientists!). In fact it turns out that nobody has ever *measured* a wave function (and most probably nobody ever will) and its common interpretation is that ψ is nothing but a tool for making calculations to predict the outcome of actual measurements. Per se ψ apparently has no physical meaning! The debate on this point has been long and passionate and it possibly will never come entirely to an end, although scientists have almost universally come to *a consensus* on the standard *minimal* interpretation given above, not least because it works so extremely well. In Chap. 13, we will discuss briefly an alternative interpretation.

Let us skip the historical debate and jump to the presently *accepted wisdom*. The wave function is considered a *probability amplitude*, its norm squared (i.e. $|\psi|^2$) being a *probability density*. If we integrate $|\psi|^2$ over a given volume, we obtain the probability of finding the particle inside that volume. More generally, we can use the wave function to evaluate the probability of obtaining a given value for the energy or the momentum (or any other physical property) of any particle anywhere.

The wave function marks an enormous difference between Quantum Mechanics and Classical Physics. The difference can be clarified with a *Gedankenexperiment* (thought experiment). Suppose you have a physical system, for which you know *all* the inherent physical laws, *all* the variables and their initial values, and *all* the boundary conditions. In classical physics, a solution of the corresponding equations yields an *exact* prediction of the results of any actual experiment (at least in principle, since, in practice, the required complete knowledge of *everything* is never possible).

Conversely, the best you can do in QM, even in principle, is to calculate the probability of one or another outcome of your experiment. In practice, if you perform only one measurement, you are totally unable to predict the result. If you repeat the experiment many times, you will find that the individual results group into a pattern corresponding to the probability distribution obtained from the wave function. The prediction capability of QM is thus restricted to probabilities, just as in a lottery, in which buying more and more tickets increases your probability of winning, but does not give you a certainty (unless you buy all tickets!). There is, however, a *negative determinism*, since the equations of quantum mechanics allow you to identify with certainty what results *cannot* come out of the experiment.

This feature of quantum mechanics is certainly disconcerting for us today, but it was even more so for the physicists of the early decades of the past century. In fact the ambition of physics at that time was to progressively gain control of *all* the physical reality. Everything was assumed to be in principle predictable, provided we could acquire the necessary knowledge. Also Einstein was completely unhappy with QM, even though he had contributed to its birth. He used to say that he could not believe that *God plays dice in order to determine how the world evolves*[10] and maintained that something had to be missing from the theory. He kept this firm belief until the end of his life. However nothing seems to be missing, and QM works so well that today most scientists live happily with it and avoid asking embarrassing questions. A pinch of pragmatism and conformism has always accompanied the development of science (as well as of other human endeavours).

8.5 Quantum Field Theory

In the previous section we have outlined the essentials of QM in the study of the interactions of single particles with external *fields,* which are not affected by the particles themselves. However, for a complete description of the real world, we must take into account that fields have *sources*, which are themselves particles. The general problem treats a plurality of particles mutually interacting via some field. But each particle is related to, or even *is* (whatever that means) a field, represented by its wave function, and generates or carries a field depending on its *electric charge* and/or other properties, such as magnetic pole strength, spin, *colour* and *flavour.*[11]

Ultimately the general scenario includes a number of fields interacting with each other through quantized mechanisms. Here the term '*fields*' describes entities continuously occupying extended regions of space (and time). Their interactions serve to identify and somehow localize the 'quanta' of the fields. The quanta are either material particles (e.g. electrons, neutrinos, quarks...) or interaction packets (e.g. photons, gauge bosons, gluons...).[12]

Here we do not even try to outline *quantum field theory*, which is the context where the above topics are described, since it requires highly sophisticated mathematics. For our purposes it suffices to remark that scientists have been able to deduce a unified[13] and consistent conceptual framework for the description of physical phenomena at the subatomic scale and at high energies that is in accord with experimental evidence.[14] It has been a long abstraction process beginning at

---

[10]See citation at the head of this Chapter.

[11]Of course these 'colours' and 'flavours' have nothing to do with the usual human perceptions, but are just conventional names for properties associated with subnuclear interactions. Their meanings and relevance will be explained in Chap. 10.

[12]We will discuss these somewhat esoteric particles in Chap. 10.

[13]Not entirely, as we shall see ahead.

[14]Actually this is not entirely true as we shall see when discussing the mass of neutrinos.

the time of Werner Heisenberg and Paul Dirac, continued during Richard Feynman's time and still going on today.

Generalizing what we have already discussed for the single particle case, we may remark that a multi-particle system is characterized by its possible *states* (i.e. general configurations including all physical parameters). These states are represented in an abstract space. The interactions are expressed and the measurable values of the relevant physical quantities are obtained by the action of appropriate *operators* (i.e. symbols, such as the square root or the derivative or more sophisticated operations, that affect the wave function). The properties of the state space and the way the operators work are expressed in terms of their generalized symmetries and *algebras*. Here *algebra* refers to a branch of mathematics facilitating the manipulation of non-numerical objects. There are many algebras besides the elementary algebra that we all learned in high school.

As the reader can see, everything looks extremely abstract, and in fact it is, but it works. In the case of quantum electrodynamics (QED), the agreement between theory and experiment is superlative[15]; nevertheless some conceptual problems are still open. Richard Feynman, whom we already quoted earlier and is one of the fathers of modern QED, referred to a fundamental step in the theory (the 'renormalization') by calling it a "*dippy process*" [3]. In truth, Quantum mechanics, as well as Relativity, includes a number of features that lie outside the boundaries of our common intuition and consequently seem to us paradoxical. In the remaining sections of this Chapter we will introduce the reader to several of them.

8.6   The Uncertainty Principle

The "*uncertainty*" or "Heisenberg" principle states that it is impossible to measure simultaneously and with arbitrary precision both the position and the momentum of a particle. Roughly speaking, we cannot know at the same time the position and the velocity of a particle. The product of the uncertainties (errors), with which these two *conjugated* variables may be ascertained, cannot be less than $\hbar/2$, an extremely small but nevertheless non-vanishing number. Any real measurement is subject to some inaccuracy (depending on the techniques employed and characteristics of the set-up[16]), but let us assume that you were smart enough to succeed in finding the *exact* position of a particle (i.e. zero uncertainty in its location). Then the Heisenberg principle would tell you that the uncertainty over its momentum (or, in practice, its velocity) must be infinite, lest the product of the two uncertainties vanishes. In other words, you would not be able to know *anything at all* about its momentum (or velocity). The reverse also holds true: if you know *exactly* the momentum of a particle, then you have no idea of where the particle is.

---

[15]We have commented already on the accuracy of QED in Chap. 3.

[16]We have discussed this type of experimental error in Chap. 3.

Momentum and position are not the only *conjugated* variables, to which the Heisenberg principle applies. The time of interaction and the amount of exchanged energy are another example of such a pair. But if the effect is so small, why should we worry about it? First, because its consequences are extremely important, even from a macroscopic point of view. And second, because the principle challenges our innate epistemology: where else are we told that there are *intrinsic* limits to the knowledge that we can acquire of the world, even if we avail ourselves of better and better experimental tools?

Technically, this bizarre property of quantum objects (i.e. objects for which QM is relevant) is an immediate consequence of the fact that the operators corresponding to two conjugated variables, such as position and momentum, applied one after the other to the wave function of the particle, produce different results according to the order of their application. In other words, the two operators *do not commute.*[17] The difference between the results obtained in the two cases is proportional to the $\hbar$ constant multiplied by the wave function of the particle. Of course this explanation may be satisfactory for theorists and mathematicians, but much less so for everybody else.

We can try here to give an intuitive analogue of Heisenberg's principle to show that it is not as outlandish as it may appear at first sight. The principle is in fact related to the nature and properties of a wave. Suppose you are facing the surface of an ocean which is stimulated by a quiet and regular wave field. As far as the eye can see you observe perfectly regular ups and downs of the surface. Mathematically this is a pure harmonic wave. The propagation velocity is exactly defined, but where is the wave? Is it everywhere? From a classical viewpoint the question is meaningless.

Now consider what happens when you beat the surface of the water with an oar. You produce a wave, or a perturbation that propagates like a wave, but is quite different from the pure harmonic wave of the first example. Now the perturbed area is small and, although there are oscillations, they are not all equal and do not extend for an infinite range on the sea surface. We can say that the wave is *localised*, but now the propagation velocity is no longer well defined. The '*group velocity*' (the technical name used in this case) somehow depends on the shape of the *wave packet* (the travelling perturbed patch of the surface).[18] Summing up: the more you localise the wave system, the less precisely you know its propagation velocity. The quantum mechanical situation is certainly more subtle, and furthermore it invokes Planck's constant, but the naïve example above should at least convey the flavour of the effect.

---

[17]As an example, the multiplication operation in ordinary algebra commutes because $a \times b = b \times a$. However, if a and b are non-symmetric matrices and not ordinary numbers, the products $a \times b$ and $b \times a$ are different and the matrix multiplication operator is said to be non-commutative. For a simpler example $\sqrt{\log x}$ is different from $\log\sqrt{x}$.

[18]In the example the problem is related to the fact that the different harmonic waves, which superpose to form the propagating perturbation, travel with different velocities depending on each one's frequency: we say that the medium (the sea water surface, in this case) is dispersive.

8.7 Superluminal Phase Waves

Although QM has been developed in accordance with special relativity, some difficulties arise when one tries to reconcile the two theories. As we have seen in Chap. 7, a fundamental tenet of special relativity is that the speed of light c is the same for all inertial observers and that *no physical object can travel faster than c*. We have also seen that in Quantum Mechanics each particle is associated with a wave function and may be described in terms of simple harmonic waves: i.e. sines and cosines. A harmonic wave propagates with a velocity, called *phase velocity*, at which the crests (or the valleys) of the wave are traveling. The phase velocity is given by the product of the frequency times the wavelength. Since electromagnetic waves can be described as streams of photons, their phase velocity is given by the ratio between the energy of the photon (which is proportional to the frequency of the wave) and its momentum (which is inversely proportional to the wavelength). As a result, the speed of the photon turns out to be exactly c.

Let us apply the same argument to a free electron moving at a given speed V (lower of course than c) and with an associated quantum wave. The phase velocity is the ratio between the total relativistic energy of the particle mc^2 and its momentum mV. The result is larger than c for a subluminal particle (V less than c), which is consequently associated with a superluminal wave (i.e. phase velocity greater than c)! This result stresses the non-physical nature of the wave function, even if it is used to evaluate physical quantities.

If we consider a real electron, which occupies some finite portion of space, we describe it quantum-mechanically as a *wave packet*, i.e. as a superposition of ideal harmonic plane waves each having a different (superluminal) phase velocity. Looking for the *group velocity* of the wave packet, we correctly get the subluminal V, the physical velocity of the particle.

8.8 Collapse of the Wave Function and Multiple Universes

Imagine an experiment in Quantum Mechanics in which an electron is fired against a very narrow slit in a wall. The wave function describing the electron before it arrives at the wall is a plane wave. Upon hitting the wall, the electron waves are diffracted by the slit in the same manner as light waves would be. After passing through the slit, the wave opens into a fan-like shape and its amplitude is modulated in alternating maxima and minima across the directions diverging from the slit. Remember that the meaning we have given to the wave function is that of a probability amplitude. The square of the amplitude is the probability density for finding the particle at any given position.

Now suppose we have another wall covered with particle detectors at some distance beyond the slit. One of these detectors registers the arrival of an electron. We know that there is only one electron, so only one counter will detect it. However just before that, the wave function had non-zero values almost everywhere, including at the location of many other detectors. We can infer that at the very moment of the interaction of the electron with one of the counters, i.e. when its position becomes precisely defined, its wave function must *instantaneously* shrink to the size of that detector and drop to zero everywhere else, e.g. at the sites of the other detectors. This is called the *collapse* of the wave function.

Once again the behaviour we have described is rather strange; it also contradicts relativity, since it requires that some kind of information travels at *infinite* speed from the position where the electron has been detected to the location of the remaining counters. The amplitude of the wave function must drop to zero *simultaneously* everywhere, but in Relativity simultaneity cannot be absolute, since it depends on the reference frame used by the observer.

This issue of the collapse of the wave function, triggered by some physical interaction has been debated at length, without finding a consistent interpretation. Once more we may stress the non-physical nature of the wave function, which plays the role of a mere mathematical tool used to compute actual physical quantities, or the probability of finding quantum objects at a given position.

However, the probabilistic interpretation, if taken literally, can lead us to paradoxical, though not inconsistent consequences. A probability density cannot abruptly jump from one value to another or, in particular, become suddenly zero. It would be consistent to assume that all possible evolutions in time ('histories'), endowed with a finite likelihood, are equally real. According to this viewpoint, each interaction is a bifurcation point of reality: the whole universe splits into as many *parallel universes* as there are possible evolutions stemming out of the interaction.

We repeat that this interpretation, proposed by Hugh Everett in 1973 in his Ph.D. thesis at Princeton, is not logically inconsistent, although it looks as though it might be (and in fact it has been) the subject of science fiction stories. It is called *Many Worlds Interpretation*. Taking this idea literally, all possible evolutions of the universe exist in parallel. There is a world in which the Romans discovered and colonized America in the first century AD; another in which Napoleon was beaten at Austerlitz; another in which Hitler won the Second World War...[19] And there exists an infinite number of copies of myself: one having preferred poetry to physics[20]; another who is managing a pizzeria... and so on.

---

[19]This is the theme of the novel *Fatherland* by Robert Harris (Hutchinson, 1992).

[20]Yea, all which it inherit—shall dissolve, And like this insubstantial pageant faded, Leave not a rack behind. We are *such stuff*, As *dreams are made* on. William Shakespeare: *The Tempest*, Act 4, Scene 1.

8.9   Entanglement and Superluminal Correlations

When we consider waves, we can of course exploit all the properties of these waves. In particular, we have already mentioned that even after the superposition (i.e. linear combination) of different waves with different parameters (i.e. different wavelength and frequency), one still has a wave. Consider now an experiment where a pair of particles is emitted with some global constraint, for example that the total spin of the particles is zero. Zero total spin in the case of a pair of electrons means that the individual spins of the two particles are opposite to each other. (Remember that the spin of an electron, in units of $\hbar$, is ½). According to QM the state of the system is represented by a wave function which is a linear combination of the wave functions corresponding to two free electrons. The combination is not arbitrary but is so made that the constraint of zero total spin is guaranteed. The global wave function nonetheless extends everywhere throughout space (and time).

Suppose then that somewhere one of the two electrons interacts with a physical device (e.g. a filter) that fixes the axis of its spin. 'Immediately' (or, if you prefer, simultaneously) the other apparently free electron, no matter how far away it is, will orient its own spin axis to maintain the total spin of the pair equal to zero. This is because the pair of electrons is described by the one global wave function that was initially generated at the source. This link between distant quantum objects is called *entanglement*. The implication of entanglement between wave functions is that once again there must exist some information[21] producing a physical effect that travels at infinite speed.

Quantum Entanglement was raised by Einstein and others in 1935 as an example of a "spooky action at a distance" that contradicted the theory of relativity. He argued that this paradox was evidence of incompleteness in the extant formulation of QM. The violation of the principle of relativity may better be understood by considering an ideal experiment made with photons rather than electrons or other particles. Imagine we have a source where pairs of photons are produced with zero total momentum and opposite circular polarizations. In practice the two photons fly apart in opposite directions and the electric field of both rotates clockwise with respect to the propagation direction: for an observer at the source the two electric fields rotate in opposite directions so that the pair is unpolarised. Before starting the test, two experimenters agree on the kind of measurement to perform and decide to check the polarisation of each photon with a quarter-wave-plate, orienting the optical axes of the two quarter-wave-plates perpendicular to each other. These orientations can be established by using the fixed stars as a reference. After equipping themselves with the appropriate devices, the two experimenters separate and reach opposite positions along the expected path of the photons.

During the test, the two experimenters do not communicate. Once the source begins to emit pairs of photons (suppose for simplicity this happens at a regular

---

[21]One needs to be cautious with the term 'information', which is used somewhat differently in QM to ordinary usage.

interval), the two experimenters record the photons passing through their quarter-wave-plate, then after a previously arranged time they send the sequence of the recorded events to a third observer positioned elsewhere. The quantum entanglement implies that when one of the photons crosses a quarter-wave-plate its wave function collapses from circular to linear polarisation along the optical axis of the plate, but the collapse of the wave function *immediately* involves the other photon, however far away it is. The consequence is that the second photon will be stopped by the other plate, whose optical axis is, by previous agreement, perpendicular to the former. The resulting sequence for successive pairs will show an anticorrelation: when one observer counts a photon the other will not, and vice versa.

Where is the conflict with relativity? We may imagine that the three observers are all at rest with respect to each other so that in their common reference frame, which event comes first (be it on the right or on the left) is well defined. However the two events happening at the two wave-plates are separated by a space-like interval, i.e. they are not causally connected, since to go from one to the other before the latter's detection event requires a velocity higher than the speed of light. We saw in Chap. 7 that in a situation like this, the time ordering of the events depends on the motion of our third observer, i.e. whether she/he is moving with respect to the other two. The ordering can be inverted simply by changing the relative velocity. However, the recorded sequence is unique, even though which photon is causing the collapse and which is undergoing it should be observer dependent.

A number of experiments over the past several decades has confirmed that entanglement is no idle fantasy, but an observable physical phenomenon, although transmission of classical information at faster-than-light speeds is still not possible [4]. Research is currently underway into the use of entanglement in computing and communication applications. We will discuss some of these applications, and further philosophical implications in Chap. 13.

8.9.1 Macroscopic and Microscopic

The strange physical phenomena that we have discussed so far belong, more or less, to the atomic scale and are not expected to be observable at the macroscopic level. In the beginning of QM, this difference led to distinctions being drawn between "the observer" (macroscopic and governed by classical mechanics) and "the observed" (subject to QM at the quantum scale). This distinction is however rather arbitrary. After all, every observer is made up of an enormous number of quantum objects. Is the macroscopic un-observability of quantum phenomena simply related to the smallness of the effect when the masses involved are so large? In other words, is it only a quantitative problem? Remember, however, that the classical mechanics we use to describe macroscopic phenomena is deterministic, whereas QM is probabilistic.

A possible, although not entirely satisfactory solution for the observer/observed dichotomy involves the *decoherence* of the wave functions. Every large object, such as ourselves, is represented by a macroscopic wave function made up of the linear combination of an enormous number of elementary wave functions associated with every single particle of it. Now when many waves superpose, a crucial role is played by the phase of each one of them. If the elementary waves are coherent, which means that their relative delays with respect to each other remain constant in time, the superposition still displays the properties of a wave, i.e. it interferes, and undergoes diffraction. However, if the elementary waves are *incoherent*, i.e. the relative delays change randomly in time, the superposition produces physical effects which are just averages of the individual effects, and such an average coincides with the expected classical behaviour. In a macroscopic object, the coherence is therefore lost.

An analogy is the light emitted from a tungsten bulb, which comes from many different atomic sources randomly distributed along the filament, and as a consequence is incoherent. The light from a laser comes from atoms which are stimulated to emit their light in phase with each other. Laser light is therefore coherent.

8.10 QM and Cats

There is a famous paradoxical *Gedankenexperiment* suggested by Erwin Schrödinger, criticizing the quantum entanglement proposed by Heisenberg and Bohr. Suppose you have a steel box and insert into it a device containing a limited amount of some radioactive substance. The quantity and type of the radionuclide are such that, in the time span of an hour, there are equal probabilities of a radioactive decay occurring, or not occurring. Now if the decay does occur, a Geiger counter detects it and triggers a mechanism breaking a phial containing cyanide. The crucial point is that at the beginning of our experiment, ignoring the protests of animal-rights activists, we had also introduced into the box a cat.[22]

After one hour we open the box: if a radioactive decay has taken place, we will find a dead cat; otherwise it will be alive. From the quantum mechanical viewpoint applied both to the radioactive nucleus and to the cat, we may say that we have two entangled states of the system: (1) an excited nucleus and a living cat; (2) a nucleus in its ground state and a dead cat. Continuing with the QM interpretation of the situation, we recall that wave functions allow for linear superpositions, and the wave function *before* an observation expresses and contains all possible states of the system. It is the observation that selects one of the states, eliminating all the others, through the collapse of the wave function.

---

[22]We hasten to emphasise that this is a thought experiment. No real cat, living or dead, or any other animal for that matter, needs to be involved in carrying out this exercise.

In the light of the above, what is the condition of the cat *before* we open the box, and in the absence of any other information on the progress of the radioactive decay? Classically we would simply say that we do not know whether the animal is dead or alive, but that it actually is in one of these two conditions. In the QM interpretation, our quantum cat is in a superposition of death and life. It is our action of opening the box that will cause the collapse of its wave function and leave it in one or the other of the two states, i.e. dead or alive.

Of course, the cat being a macroscopic object, we may call in the decoherence mechanism and conclude that it does not behave as a quantum particle, but where is the border between the macroscopic classical and the microscopic quantum domains? There is also a problem concerning the roles of the observer and the observed. If I am the observer, the cat is in a superposition of life and death. However, why shouldn't the cat be the observer, in which case I am in a superposition of sorrow for its death and relief for its survival, as long as the box is closed?

There are a number of interpretations of the Schrödinger cat experiment, some involving different universes populated with live and dead cats. Suffice it to say here that there are still many open questions relating to the interface between the micro quantum world of atoms and sub-atomic particles, and the macro world in which we live our daily lives.

In this Chapter we have presented some of the more bizarre concepts of Quantum Mechanics. It should not be thought, however, that QM is the plaything of daydreaming physicists in the ivy-covered towers of ancient universities. QM is essential to the understanding of all phenomena that involve small particles (i.e. electrons, nuclei, atoms, molecules, etc.). Its use in the world of atoms is as much routine as is the theory of Newton in the mechanics of our everyday life. In the next Chapter, we will encounter some of these applications. QM also raises a number of philosophical questions, which we shall address in the final Chapter of this book.

References

1. R. Feynman, *Probability and Uncertainty—the Quantum Mechanical View of Nature*, chapter 6, p. 129
2. A. Dante, *The Divine Comedy* (Book I, Hell, Canto 3), translated by Clive James, Picador
3. R. Feynman, *QED: The Strange Theory of Light and Matter*. (Princeton University, 1965) p. 128
4. R. Penrose, *The Road to Reality*. (Jonathan Cape, 2004) p. 578

Chapter 9
Atomic and Nuclear Physics

Nothing exists except atoms and empty space; everything else is opinion.

Democritus of Abdera

Abstract The ancient Greek idea of atoms as the basic building blocks of matter is consolidated by the work of Dalton, Lavoisier, and Mendeleev. Rutherford shows that the atom has a nucleus, and the new science of atomic spectroscopy leads to the Bohr-Rutherford model of the atom and the "old" quantum theory. Modern quantum mechanics is used to explain the distribution of electrons about the atoms. The nucleus itself is found to have an internal structure, and the concepts of nuclear fission, nuclear fusion, nuclear forces and nuclear models are explained. The synthesis of the elements in the stars is discussed.

9.1 Early Days

The insightful quotation heading this chapter is by Democritus of Abdera, a Greek philosopher who lived in the fifth Century B.C. (c. 470 B.C.—c. 380 B.C.) He continued his explanation: *"The worlds are unlimited. They come into being and perish. Nothing can come into being from that which is not, nor pass away into that which is not. Further, the atoms are unlimited in size and number, and they are borne along in the whole universe in a vortex, and thereby generate all composite things—fire, water, air, earth. For even these are conglomerations of given atoms. And it is because of their solidarity that these atoms are impassive and unalterable."*

This idea, that everything in the universe is composed of small, immutable particles called atoms that are in constant motion in a vacuum and collide with each other, encapsulates the basic principle of modern atomic theory. However, Democritus did not carry the same level of prestige and influence among his peers as Aristotle, and when the great man rejected his rival's hypothesis, atomic theory

© Springer International Publishing Switzerland 2016
R. Barrett et al., *Physics: The Ultimate Adventure*, Undergraduate Lecture Notes in Physics, DOI 10.1007/978-3-319-31691-8_9

was ignored for the next two millennia. Such is the unfortunate outcome of an over-developed respect for authority.

It was not until the early nineteenth century that John Dalton, a British Chemist investigating the proportions of elements present in chemical reactions, resurrected the concept of atoms. Extending earlier work of Joseph Proust and Antoine Lavoisier, he proposed in 1808 his law of multiple proportions: "*If two elements form more than one compound between them, then the ratios of the masses of the second element which combine with a fixed mass of the first element will be ratios of small whole numbers.*" For example, if we consider the two carbon compounds, carbon monoxide and carbon dioxide, the mass of oxygen combining with (say) 100 g of carbon in carbon dioxide is twice the value that it is in carbon monoxide. This is easy to understand when we realise, as Dalton did, that a molecule of carbon monoxide contains one carbon atom and one oxygen atom, while a molecule of carbon dioxide contains one carbon atom and two oxygen atoms.

Dalton postulated that elements (substances that cannot be decomposed into simpler constituents, in contradistinction to chemical compounds, which can) consist of small, immutable particles called atoms. Atoms of a given element are identical, but those of different elements differ. Molecules of chemical compounds are formed in chemical reactions from combinations of simple whole numbers of the atoms from different elements. These concepts led to the blossoming of the science of chemistry.

After the demonstration by Newton in 1666 that white light could be split by a prism into the colours of the rainbow, other researchers applied his technique to investigate the light emitted from various sources, both terrestrial and celestial, giving birth to a field of study now known as spectroscopy, (see Chap. 8). By the middle of the nineteenth century Gustav Kirchhoff and Robert Bunsen, building on the work of Wheatstone, Stokes, Foucault, Thomson and others, had established that emission spectra often contained lines that were a characteristic of the element constituting the source. Indeed, an element not yet seen on earth at that time was postulated from a series of unexplained lines in the solar spectrum. It was called helium, in recognition of its presence in the sun.[1]

Kirchhoff gathered the results of his research into three important laws:

1. A hot solid object produces light with a continuous spectrum.
2. A hot tenuous gas produces light with coloured spectral lines at discrete wavelengths characteristic of the source.
3. A cool tenuous gas surrounding a hot solid source shows the continuous spectrum of the solid object, but with gaps at discrete wavelengths which are characteristic of the surrounding gas. Earlier Joseph von Fraunhofer had observed lines of this type, which still bear his name, in the spectrum from the sun.

In 1885 Johann Balmer, through a careful study of the line spectrum of hydrogen, the lightest of the atoms, discovered a simple empirical formula

---

[1]Helios, in Greek.

involving integers for the frequencies of four of the spectral lines. A few years later Johannes Rydberg discovered a similar formula for another series of lines in the hydrogen spectrum. These formulae were very accurate, but they were completely empirical. A theoretical understanding of the origin of spectral lines would have to wait for the revolution in modern physics in the 20th century.

Dalton's work on the law of multiple proportions led to estimates of the relative weights of atoms by measurement of the weight gains and losses as the ratios of atoms are changed in chemical reactions. The atomic weight of hydrogen was set arbitrarily at unity.[2] Chemists began to notice that arranging atoms in order of their atomic weight reveals periodicities in the chemical properties of the elements. For instance, fluorine, chlorine, bromine and iodine are very active chemicals, while helium, neon, krypton, argon, xenon and radon are chemically inert.[3]

In 1869 Russian chemist Dimitri Mendeleev started the development of what is now called "the periodic table". He arranged the elements in order of atomic mass, placing those with similar properties in groups. It became clear that some elements were missing, and he left gaps in his table to accommodate their future discovery. The explanation of this strange periodicity in chemical properties had also to wait until the twentieth century for understanding.

9.2   Atomic Models

Around the turn of the nineteenth century, the discovery of a number of sub-atomic particles put an end to the idea of an atom as the ultimate fundamental particle. In 1897 J.J. Thomson discovered the electron, and experiments on radioactivity revealed the existence of three distinct forms of emitted particles. These were known as alpha, beta and gamma particles, and appeared to result from the spontaneous disintegration of certain types of atoms.

As electrons, which are the basis of electricity, carry negative charges, the other constituents of the electrically neutral atom must be positively charged. The question was: how are these positive and negative charges distributed within the atom. Thomson proposed a model which became known as the plum pudding model[4] in which the electrons are dotted through an amorphous, positively-charged material in the same fashion that dates and raisins are dotted through the traditional English Christmas dessert.

In a series of experiments between 1908 and 1913 by Hans Geiger and Ernest Marsden under the direction of Ernest Rutherford at the Physical Laboratories of the University of Manchester, Thomson's model was disproved in a dramatic and

---

[2]Strictly speaking these quantities should be called atomic masses, and that is the current practice. However in 19th century chemistry the definitions were not so precise.

[3]Helium was not discovered on earth until the end of the nineteenth century.

[4]In the United States of America it apparently became known as the blueberry muffin model.

convincing fashion. Alpha particles, the heaviest of the three forms of radioactive particles and positively charged, were allowed to impinge on a thin foil of gold. The alpha particles scattered from the atoms of gold in the foil. If the electrons in the gold atom had been distributed through a positively charged medium as Thomson's model maintained, the alpha particles would have been only slightly deflected from their incident paths. However, many alpha particles passed through the foil unscattered, while some were scattered by very large angles, essentially being reflected back on their tracks, as illustrated in Fig. 9.1.

In 1911, Rutherford carried out a mathematical analysis of the experimental results which showed that the observed scattering pattern could only arise from an atom where all of the positive charge was concentrated at the centre. In addition, the fact that a few alpha particles underwent scattering while most simply passed through the foil unimpeded implied that this positive charge must be concentrated in a very small volume, with a diameter about 1/3000th of the atomic diameter. This heavy, positively-charged atomic core was called the "nucleus" of the atom, and Rutherford's 1911 paper heralded the birth of what is now known as nuclear physics.

But how were the electrons distributed about the positively charged nucleus? Why did the negatively-charged electrons not become drawn into the nucleus by the attractive force between the unlike electric charges? The solar system, with its heavy central sun orbited by a number of planets was raised as a possible analogy to the atomic system. The planets do not fall into the sun under the influence of gravitational attraction because they are moving at a speed fast enough to allow them to fall around, rather than into, the sun. Any faster orbital speed would see them fly off into space. Surely some similar process must be keeping the electrons in orbit about the nucleus.

The astute reader will realise from what we have discussed in earlier chapters that this model has a fundamental difficulty. Any body in orbit about another heavy body is undergoing an acceleration directed inward towards the heavy body. However an electron, unlike a planet carries an electric charge. From the work of Maxwell we know that an accelerating electric charge emits energy in the form of electromagnetic radiation. This energy must be found from the kinetic and potential

Fig. 9.1 Paths followed by alpha particles scattered from a gold nucleus. Only those particles passing close to the nucleus are scattered significantly

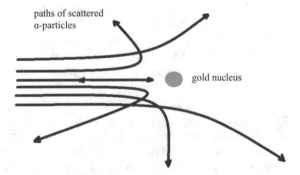

energy of the orbiting electron, with the result that the electron will slow down and fall into the nucleus. Stable planetary orbits of electrons around the nucleus are not possible according to Maxwell's theory of electromagnetism.

There is another problem with the very concept of an orbit for an electron. To describe an orbit we need to know simultaneously the position and velocity of the electron as a function of time. From Heisenberg's Uncertainty Principle, outlined in Chap. 8, we know that this is possible only to certain limits. Of course, in 1911 Heisenberg was only ten years old, so this problem was not an issue raised at the time. We will come back to this point later.

9.3   The Bohr-Rutherford Model of the Atom

A solution was proposed by Niels Bohr in 1913, in which he arbitrarily postulated that a series of stable orbits existed that violated Maxwell's theory. Electrons could stay in such orbits indefinitely, but when they received an excitation, e.g. by a collision with another electron passing by, they could be excited to a higher, more energetic orbit. When they fell back to their original orbit, they emitted a burst of radiation, or photon, with an energy equal to the energy difference between the two orbits. The frequency of this radiation was given by Planck's formula (see Chap. 8). The situation is illustrated in Fig. 9.2 for the hydrogen atom, which has only one electron.

To obtain the frequency of the emitted radiation, a knowledge of the energies of the stable orbits is required. It was known from De Broglie that an electron possessed wavelike qualities, and the wavelength λ of the electron wave was given by:

$$\lambda = \frac{h}{mV}$$

where h is the Planck constant, m is the electron mass, and V is its velocity.

Fig. 9.2 According to the Bohr model, the transition of an electron from an outer to an inner orbit in the hydrogen atom results in the emission of a photon with an energy given by Planck's formula: $E = h\nu$

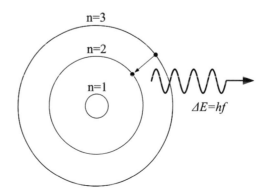

The energy levels in the hydrogen atom were obtained by Bohr by assuming that the circumference of the electron orbit is a multiple of the de Broglie wavelength λ, i.e.,

$$2\pi r = n\lambda,$$

where r is the radius of the orbit, and n is an integer denoted as the principal quantum number. As the angular momentum L of the electron in an orbit is:

$$mVr = \frac{nh}{2\pi}$$

we see that the angular momentum of the electron is also quantised, i.e., it can only have particular discrete values.

Despite its arbitrariness, Bohr's theory had some spectacular successes. It predicted the frequencies of the Balmer and Rydberg spectral lines precisely, and also the frequencies of lines that had not yet been observed.

More complex atoms have several electrons orbiting about the nucleus. In this case, the electrons interact with each other, as well as the nucleus, and we have a many-body problem, which as we have seen earlier, is insoluble. The increased positive charge of the nucleus in these heavier atoms increases the strength of the electric potential in which the electrons orbit, and as a consequence increases the frequency of the electronic radiation when a quantum transition takes place, particularly between the deep-lying orbits. This radiation is in the form of x-rays, rather than the visible and UV frequencies of the Balmer and Rydberg hydrogen lines.

One might naively expect all electrons to radiate energy in the form of photons until they occupied the lowest orbit, or shell, around the nucleus. However, the number of electrons that are allowed in each shell is a function of the angular momentum associated with that shell. Two electrons are allowed in the innermost shell, eight in the second, and higher numbers in outer shells.[5] The number of electrons in the outermost shell determines the chemical properties, e.g. the valency, of an element, and leads to the familiar Periodic Table of the Elements.

The spectacular success of the Bohr model for the explanation of atomic spectroscopy is marred by the arbitrariness of its assumptions. Why should some orbiting electrons not emit radiation as required by Maxwell's laws, while others do? It was clear, even to Bohr, that his theory was only a stepping stone to a better theory. That better theory is quantum mechanics.

---

[5]This is an application of an exclusion principle formulated by Wolfgang Pauli in 1925.

9.4   The Quantum Mechanical Picture

As we have seen in Chap. 8, ignoring relativistic effects, the motion of a particle is described by Schrödinger's wave equation, which returns a probabilistic determination of where that particle is most likely to be found. The electron in a hydrogen atom moves in a potential energy field that is attractive. If the energy of the electron is insufficient to escape from this field, the electron is said to be "bound". To determine the predictions of quantum mechanics for such an electron, we need to solve the Schrödinger wave equation for an electron in a potential energy field created by the positively charged nucleus. Such a task is beyond the scope of this book. However, it is possible to gain some understanding by a comparison with other, more familiar physical systems.

The motion of a string stretched between two fixed supports is also described by a wave equation. When waves pass along the string, they reflect at each end, so that the reflected wave interferes with the incoming wave. The result is a standing wave pattern. The string is at rest at each end, but along the string the vibrational amplitude of the string increases until it reaches a maximum, or antinode. The amplitude then decreases until it becomes zero (at a node) and then it begins increasing again. There may be a number of nodes and antinodes along the string.

The simplest case is where there is an antinode in the centre of the string, and nodes at each end. This is known as the fundamental mode of vibration. In a musical instrument this mode results in the lowest pitch that can be obtained from that string. The other modes of vibration, with more than one antinode, are harmonics, and the notes emitted from the string in this case are at a higher pitch than the fundamental, but related to it by a known musical interval.[6] The frequency spectrum of sound emitted by the string is not continuous, but "quantised", in a similar manner to the quantisation of atomic spectra.

The description of the vibration of the surface of a drum requires the solution of a two dimensional wave equation. In this case there are nodal lines on the surface of the drum where the vibrations are minimised. The higher order vibrations of the surface are not harmonic in a musical sense because their frequencies are not integrally related to the fundamental frequency of vibration, as is the case with the one-dimensional string.

Three dimensional standing wave patterns can also occur, for instance, in the microwave radiation inside a microwave oven. To avoid having food located at a node in the standing wave pattern, where it would remain uncooked, the food is placed on a rotating tray so that it moves through antinodes as well as nodes during the cooking process.

Returning to the example of the hydrogen atom, the probability of an electron being located at a given point in space around the nucleus is obtained by solution of Schrödinger's equation, which is a wave equation not unlike that describing

---

[6]For instance, the ratio of frequencies of notes an octave apart are 2:1, for a perfect fifth: 3:2, for a perfect fourth: 4:3, for a major third: 5:4, for a minor third: 6:5.

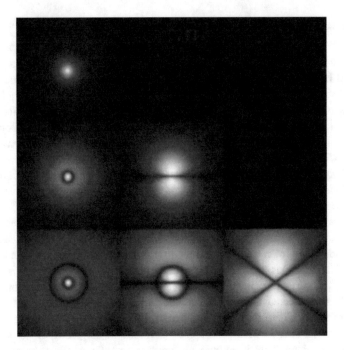

Fig. 9.3 Probability distribution of electrons in the hydrogen atom for the ground (fundamental) state (at *top left*) and selected excited states. The highest probabilities are where the image is brightest

microwaves inside a microwave oven. Only discrete energies are possible. The lowest energy solution corresponds to the fundamental mode of vibration in the three dimensional wave problem. The nucleus can be considered to be surrounded by a cloud, the density of which at a particular point is a measure of the probability of finding the electron at that point. There is no precise orbit for an electron in the quantum mechanical picture. As we have seen, such an orbit would violate Heisenberg's uncertainty principle.

A graphical depiction of the electron probability distribution for the lowest energy (or fundamental) state and several excited states of the hydrogen atom is shown in Fig. 9.3.[7] For higher energy states, fluctuations in the density corresponding to nodes and antinodes occur, and the electron has a greater probability of being located further from the nucleus.

The beauty of the quantum mechanical solution is that the discrete energy levels it calculates correspond exactly to those obtained from Bohr's arbitrary proposals.

---

[7]This image has been slightly modified from the image uploaded by Florian Marquardt to Wikimedia Commons and made available under the Creative Commons Attribution-Share Alike 3.0 Unported license.

(https://commons.wikimedia.org/wiki/File:HAtomOrbitals.png).

The shell structure of Bohr's model, which was essential for the explanation of chemical properties and the periodic table, is retained. Bohr's model is therefore superseded by a much more powerful theory, which can be applied to a myriad of atomic physics problems besides the simple hydrogen atom. It is another example of a somewhat arbitrary physical model giving way to a theory that absorbs the earlier work into its purview while offering an explanation to so much more than was possible with the limited earlier model. In this way, Kepler's laws of planetary motion were overtaken by Newton's theory of gravity, and Newton's theory itself was later absorbed into Einstein's theory of relativity.

The Bohr-Rutherford picture of the atom also raises questions as to the nature of the nucleus. How can the positively charged entities that comprise the nucleus remain together? Surely their repulsive electromagnetic forces must drive them apart. These issues will be explored in the remaining sections of this chapter.

9.5   Inside the Nucleus

The Bohr-Rutherford model of the atom may inspire the notion of an inert nucleus surrounded by a cloud of electrons, but that picture is far from the reality. It is true that the chemical properties of the elements are explained by the interactions of the electrons, most notably the outermost ones, but the nucleus itself is a vibrant quantum system. Not all nuclei are stable, i.e. persisting in their current state for geological eons. Some decay spontaneously, emitting a form of nuclear radiation to remove energy, mass and charge, thereby transforming themselves into the atomic nuclei of entirely different elements. The transformation of the elements, sought for so long by medieval alchemists, is a physical reality.

This spontaneous nuclear decay is called radioactivity, and was discovered serendipitously in 1896 by Henri Becquerel when he left uranium salts lying in a drawer on a photographic plate. Geiger and Marsden employed one form of radioactivity, alpha radiation, as the source of the particles in their experiment that led Rutherford to postulate the existence of the atomic nucleus in 1911. As we can see, products of nuclear decay were actually discovered prior to the formulation of the concept of an atomic nucleus.

Once it was clear that a hydrogen atom is comprised of a central, positively-charged nucleus with an orbiting electron (in the Bohr-Rutherford model), the natural extension for heavier atoms was to presume that their nuclei contained multiples of hydrogen nuclei, or protons as Rutherford called them. To maintain electrical neutrality, an atom would need the same number of protons in its core as it has electrons surrounding it. One would therefore expect the atomic weight of an element to be a multiple n of the hydrogen atomic weight, where n is the number of electrons contained in its atom. No explanation of the origin of the force required to bind the protons together against the repulsion of their electric charges was available at this time.

A close examination of elements in the periodic table showed that the atomic weight of many elements was indeed a multiple of the hydrogen atomic weight. However this multiple was not the n described above, but its value lay in the range from 2n to 3n. To confuse the picture further, the atomic weight of many elements in no way approximated a multiple of the hydrogen atomic weight.

To explain these somewhat peculiar results, Rutherford proposed in 1920 that the atomic nucleus must be constructed from protons, i.e. hydrogen nuclei, combined with other particles that have the same mass as a proton, but no charge. He called these particles neutrons. The neutron was discovered experimentally in 1932 by James Chadwick, a student of Rutherford. The neutron became one in a line of particles, ranging from the neutron and neutrino through to the Higgs Boson, that have been predicted by theory before their experimental discovery.

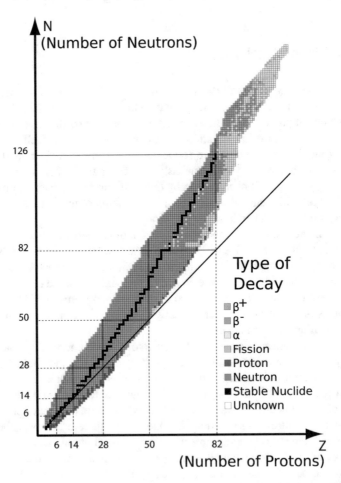

Fig. 9.4 The line of stability of nuclides. Nuclei outside the coloured areas are unstable, and not observed in nature

We now know that Rutherford's hypothesis was correct. Neutrons and protons are gathered collectively under the name "nucleons", and apart from an absence of electric charge, the neutron is very similar to the proton. It is possible for two atoms to contain the same number of protons,[8] but different numbers of neutrons. These two types of atom are said to be isotopes. The electron cloud surrounding their nuclei, and therefore the chemical properties of the atoms, depends only on the atomic number, Z. A naturally occurring element may in fact be a mixture of different isotopes, which explains why the atomic weights of some elements are not integral multiples of the hydrogen atomic weight.

We might consider a nucleus to be comprised of arbitrary numbers of neutrons and protons. However this is not the case. Figure 9.4,[9] which plots the number of protons against the number of neutrons, shows a line of stability along which all naturally occurring elements are located.

9.6   Nuclear Decay

Nuclei to either side of this line of stability will decay, emitting radioactive particles to transmute themselves into reaction products nearer to the line of stability. Nuclei with atomic numbers greater than 92 (uranium) do not exist naturally, but have been created artificially in reactors and nuclear accelerators. They are short-lived by geological time scales, although some still have half-lives of thousands of years.[10]

Nuclei with certain specific values of atomic number Z or neutron number N appear to be more stable than the other nuclei around them. These values are known somewhat unimaginatively as "magic numbers", and are 2, 8, 20, 28, 50, 82, and 126. Nuclei with both Z and N equal to one of these numbers are said to be "doubly magic". Examples of doubly magic nuclei are helium (Z = 2, N = 2), oxygen (Z = 8, N = 8), calcium (Z = 20, N = 20), and lead (Z = 82, N = 126).

Unstable nuclei may decay by a number of processes.

1. *Alpha decay* This was the radioactivity discovered by Becquerel, and occurs when a heavy nucleus ejects a particle made up of 2 protons and 2 neutrons, i.e. it emits the nucleus of a helium atom. Examples of alpha emitters are natural isotopes of radium, thorium and uranium, as well as the artificially created trans-uranics,[11]

---

[8]The number of protons in an atom is known as its "atomic number, Z." The sum of the atomic number and the number of neutrons, N, in an atom is its "mass number", A.

[9]This image was obtained from Wikimedia Commons and made available under the Creative Commons Attribution-Share Alike 3.0 Unported, 2.5 Generic, 2.0 Generic and 1.0 Generic license.
(https://commons.wikimedia.org/wiki/File:Table_isotopes_en.svg).

[10]The half-life of a radioactive isotope is the time required for half of its mass to undergo radioactive decay.

[11]These are nuclei with Z > 92 which have been created in reactors or with nuclear accelerators.

americium and plutonium. Alpha emitters are very dangerous if ingested, but they do not penetrate far through matter, and are easily shielded.

2. *Beta minus decay* In this form of decay, the nucleus ejects an electron. It is tempting to think that the electron may arise from the electron cloud surrounding the nucleus. However, this is not the case. The emitted electron arises when a neutron in the nucleus converts itself into a proton, and ejects an electron to balance the electric charge. However, as will be discussed in Chap. 10, such a process would not conserve angular momentum, and another uncharged particle is also emitted. To conserve energy this particle, now called an antineutrino, must be almost massless.

   The process can be represented by $n \rightarrow p + e^- + \bar{v}$ where $\bar{v}$ is an antineutrino, and n, p and e^- are a neutron, a proton and an electron respectively. The effect is to change a nucleus having too many neutrons, i.e. above the line of stability in Fig. 9.4, into one closer to, or on the line.

3. *Beta plus decay* This is analogous to beta minus decay, except that it occurs for nuclei below the line of stability, and moves them nearer to the line. In this process a positively charged electron, or positron, is emitted. A positron is yet another particle predicted by theory before it was discovered experimentally.

   The process is represented by $p \rightarrow n + e^+ + v$ where e^+ is a positron and v is a neutrino.

4. *Gamma radiation* Understandably, the sudden transmutation of one of its fundamental constituents may leave a nucleus in an excited state, much as occurs in an atom when one of its electrons is excited. Just as the excited electron loses energy in the form of electromagnetic radiation to fall into a lower orbit in the Bohr model, so will the excited nucleus lose energy in the same manner. We will discuss nuclear models more in a later section. For the moment, suffice it to say that the nucleus also changes energy levels by emitting electromagnetic radiation. Because the energies involved in nuclear transitions are very much higher than in atomic transitions, the photons emitted are of a very high frequency (thousands or millions of times those involved in the hydrogen atomic transitions). These photons are highly penetrating and therefore dangerous.

5. *Fission* Fission is the name given to a process when a nucleus splits into two parts of roughly comparable mass. Spontaneous (i.e. non-induced) fission is comparatively rare, but can occur in a few heavy isotopes. Of much more relevance is induced fission, when a heavy nucleus absorbs a particle (e.g. a neutron), thereby becoming unstable and undergoing fission. An example is shown below:

$$n + {}^{235}U \rightarrow {}^{236}U \rightarrow {}^{92}Kr + {}^{142}Ba + 3n.$$

Such a reaction is strongly exothermic, producing a vast amount of energy. The energies involved in nuclear processes are six orders of magnitude above those involved in ordinary chemical reactions. This accounts for the explosive power of nuclear weapons. Also of note is that the fission, or "splitting of the atom", induced by a single neutron produces further neutrons capable of inciting fission in other neighbouring nuclei. If the mass of uranium exceeds a certain "critical mass",[12] a self-sustaining chain reaction will take place. This is the principle underlying the nuclear fission reactor where, with a few notable historical exceptions, the chain reaction has been controlled, and the fission bomb where the reaction is uncontrolled.

The reaction described above is only one example of fission. In general, a variety of reaction products are produced. The fission products may themselves decay further, emitting radioactive particles and gamma radiation until they approach the line of stability. Some fission products are radioactive with a long half-life by human standards, and therefore must be disposed of safely.

9.7   Nuclear Synthesis

The converse of nuclear decay is nuclear synthesis. The core of a nuclear reactor is a region where neutrons produced from the fission process are in abundance. If a non-fissile element is introduced into this region, its nuclei may capture some of these neutrons to produce an isotope of the element that is not found in nature. As the new isotope is likely to be unstable because of the extra neutrons its nucleus now contains, it may undergo beta minus decay, changing one or more of these neutrons into protons. This process forms a new element with higher atomic number than the original one. By this means the trans-uranic elements neptunium ($Z = 93$), plutonium ($Z = 94$) and americium ($Z = 95$) have been formed. Elements with $Z > 95$ have been created by accelerating small nuclei in a particle accelerator and colliding them with a heavy nucleus.

The plethora of elements existing in nature today were not present in the early stages of the universe. The synthesis of these elements is discussed in Chap. 11 of this book. However an early difficulty encountered in the study of the synthesis process is deserving of mention here. The only promising explanation for the synthesis of carbon-12 was through the fusion of three alpha particles. The difficulty was that the probability of three alpha particles remaining together long enough to fuse was extremely small. One of the pioneers of this field was the controversial physicist (and science fiction writer) Fred Hoyle. He proposed that this fusion would be possible only if carbon-12 possessed an excited state with an energy equal to Beryllium-8

---

[12]The critical mass of a fissile material depends on its surrounds and whether they reflect the emitted neutrons back into the material, but is of the order of ten kilograms.

(an intermediate product in the fusion process) and an alpha particle. Subsequent experiments located this excited state at the precise energy predicted by Hoyle.

Without this excited state, the synthesis of atomic nuclei would not have proceeded to carbon-12 or beyond. The universe as we know it, including us, would not exist. The existence of this excited state at the requisite energy has been attributed (controversially) by some as an example of the "anthropic principle." We will discuss the anthropic principle further in the final chapter of this book.

9.8   Nuclear Forces

It is clear that to maintain the nucleus intact against the electrostatic repulsive forces between the protons, some new attractive force must be present.[13] This force is known as the strong nuclear interaction. It acts as an attraction between all nucleons (i.e. protons and neutrons) in the nucleus. The force must be short range, because it produces no effects outside the nucleus, and it must have a repulsive core to prevent the nucleus collapsing on itself.

Nuclear dimensions are measured in femtometres (fm) where 1 fm = 10^{-15} m. (This unit is also called the Fermi in honour of the Italian nuclear physicist, Enrico Fermi.) Nuclear radii can be measured by scattering experiments similar in character to that carried out by Geiger and Marsden, and are typically ~ 5 fm, depending on the number of nucleons in the nucleus. (For comparison, the radius of the oxygen atom, including its electron shells, is about 60,000 fm.)

For distances of less than 0.7 fm, the nuclear force between two nucleons is strongly repulsive, as discussed above. At 1 fm separation, the force becomes strongly attractive. This attraction weakens with increasing separation of the nucleons until at about 2.5 fm it becomes insignificant. The strength of the nuclear force is enormous. The energies involved can be estimated by comparing the mass of a nucleus with the sum of the masses of its constituent nucleons. The nucleus is lighter than its constituents by an amount equal to the nuclear binding energy divided by c^2, where c is the velocity of light. This is an example of the use of Einstein's famous formula, $E = mc^2$ (see Chap. 7). The strength of the nuclear force is the explanation for the vast amounts of energy released in nuclear explosions.

By now we should be comfortable with the concept of a force acting at a distance between two objects. After all, we have had several centuries of familiarity with gravity and electromagnetic interactions, and several chapters of this book, to get used to the idea. The nature of the nuclear force, however, is quite different from these two well-known interactions. For a start, it is short range, whereas gravity and the electrostatic interaction fall off as $1/r^2$, where r is the separation of the two masses (in the case of gravity) and the two charges (in the case of electrostatics).

---

[13]The force of gravity is too small by many orders of magnitude to achieve this purpose.

In addition it is found to depend on the angular momentum of the two nucleons, and three-nucleon forces may also play a significant role.[14]

From its beginnings, the aim of nuclear physics has been to explain nuclear structure and nuclear reactions from the basic forces between nucleons. In 1935, Hideki Yukawa explained the nuclear force as arising from the exchange of particles called mesons, in much the same way as the electrostatic force arises from the exchange of photons (see Chap. 8). Mesons were subsequently discovered in 1947. However, the development of the Standard Model for Fundamental Particle Physics (see Chap. 10) has seen both mesons and nucleons interpreted as combinations of quarks and gluons. In this model, the nuclear force arises as a residual interaction left over from the strong forces within a nucleon that bind the nucleon together.

The difficulty of developing an understanding of the quantum structure of the nucleus should not be underestimated. Firstly, the nucleus is a many-body system. Many-body systems are inherently mathematically insoluble. However, in for example the solar system, where the nature of the basic force (gravity) is understood, and the components are well separated, computational methods leading to successively more accurate approximations have been developed. In the nuclear case, the nucleons are all strongly interacting, the nature of the inter-nucleon force is not known precisely, and this force may depend on the angular momentum of the two nucleons and the presence of other neighbouring nucleons. As Sherlock Holmes might have observed, nuclear physicists have been presented with a "classic three pipe problem."

The advances in modern computing have seen some progress in the calculation of nuclear properties from underlying nuclear forces for the lightest nuclei. However, the traditional approach to this problem has been the development of nuclear models, which still remains the only feasible approach for heavier nuclei. We will outline some of these models in the next section.

The strong nuclear force discussed above does not act on leptons, of which the commonest example is the electron. The emission of electrons and positrons in beta decay requires a further force, the weak nuclear interaction, which is the fourth of the fundamental physical forces. The others are the strong nuclear force, the electromagnetic force, and gravity. Taking the strength of the strong nuclear force as unity, the strengths of the other forces are: electromagnetic = 1/137; weak nuclear = 10^{-6}; and gravity = $\sim 6 \times 10^{-39}$. The range of the weak force is about 10^{-18} m, which is 10 % of the proton diameter and 1/1000 the range of the strong force.

In the Standard Model of fundamental particles (see Chap. 10), the weak force involves the exchange of intermediate vector bosons, W and Z, and is involved in the change of quarks from one flavour to another. It is important in all processes in nuclei that involve leptons.

---

[14]By this statement we mean that the force between two nucleons will be different if a third nucleon is in their vicinity.

As discussed in Chap. 3, the weak nuclear force is the only one of the fundamental forces that distinguishes between left- and right- handedness. For the vast
majority of physical processes, a looking-glass world is indistinguishable from the
real one. A popular thought experiment is to ponder how one might communicate
by words alone to someone in a faraway galaxy that our hearts lie on the left side of
our chests. The terms left and right are purely a matter of convention, and the
everyday physical laws have no left-right asymmetry that we could exploit. It
would seem that a solution (perhaps the only solution) would be to convey to them
instructions for carrying out an experiment that involves the weak nuclear interaction, for instance a scattering experiment using nuclear beta decay, similar to the
one performed by Madam Wu in 1956.

9.9   Nuclear Models

In Chap. 3 we have presented the concept of a mathematical model in physics as a
tool to be used when direct explanation of physical phenomena from fundamental
principles is not possible, i.e. when the situation under investigation is too complex
to be analysed directly from the basic physical laws. For instance, when studying
liquids, physicists have introduced concepts such as 'surface tension' and 'viscosity' which describe well-known effects in the properties of liquids. In principle,
these properties of liquids could be derived from the forces between the molecules,
which in turn are a product of the electromagnetic forces between the outer electrons in the atomic shells. However, such a derivation is impractical because of the
large numbers of particles present in even a small drop of liquid. Although these
large numbers preclude basic derivations, the ensemble nature of the liquid enables
statistical concepts such as viscosity and surface tension to be modelled accurately.
We have already discussed in Chap. 5 how statistical mechanics can be applied to
predict the Ideal Gas Law, and similar bulk behaviour in gases.

Whereas there are about five sextillion (5×10^{21}) atoms in a drop of water, even
the largest nucleus contains less than 250 nucleons. This number is large enough
that the derivation of nuclear structure from basic inter-nuclear forces is not possible, but small enough that statistical techniques are likely to have large uncertainties in their predicted results.

For classical liquid drops, the density of the liquid in the drop does not depend
on the size of the drop, and the heat of vaporisation of a drop is proportional to the
mass of the drop. The latter is equivalent to stating that the energy required to
remove each molecule from the drop is independent of the drop size. In the case of
the nucleus, it was discovered from the measurement of nuclear radii that the
interior mass densities inside the heavier nuclei do not vary greatly from nucleus to
nucleus. It was also found that the binding energy per nucleon is approximately
constant. There is a clear analogy between the classical liquid drop and the atomic
nucleus, and as a consequence a 'liquid drop model' was developed for the nucleus
to explain certain nuclear properties.

One use of this model was to obtain a semi-empirical mass formula to predict the nucleon binding energy. This formula contained terms, such as a surface term that can be related to the concept of a surface tension, in direct analogy to the classical liquid drop. Other terms, e.g. a Coulomb term arising from the charges carried by the protons, have no such direct analogy. Despite its simplicity, the liquid drop model is able to describe fairly accurately the masses of hundreds of nuclei from five basic empirical parameters, and is useful in exploring nuclear fission.

There are some nuclear properties that cannot be explained by the liquid drop model, no matter how much the parameters are tweaked. One of these we have already encountered: the presence of so-called 'magic numbers' of nucleons. A nucleus containing a magic number of neutrons or protons appears to be considerably more stable than its neighbours. Students of chemistry will be familiar with the inert nature of the noble gases, helium, neon, krypton, argon, xenon and radon. An explanation of the stability of these gases had been obtained in terms of atomic shells. Electrons subjected to the electric field of the nucleus arrange themselves in shells of increasing energy, according to the requirements of the Pauli Exclusion Principle and the restrictions of angular momentum conservation. In each of the noble gases, the outermost shell is completely filled, which accounts for the gas's inert properties. Perhaps some analogous shell structure might account for the magic numbers in nuclear physics. These ideas led to the development of the nuclear shell model.

In the atomic case the electrons are free to move in the electromagnetic potential well created by the heavy positively-charged nucleus. However, inside the nucleus, there is no comparable central core. The nucleons move inside a potential well that is created by the sum of the attractive forces of all the other nucleons. The depth of the well is constant inside the nucleus since the nucleons are distributed uniformly in this region. The nucleons occupy the energy levels of this potential well in such a way as to minimise the total energy, while at the same time not violating the Pauli Exclusion Principle.

The 'shell model of the nucleus' involves the solution of the Schrödinger wave equation for a nucleon in a central potential well to obtain the allowed energy levels for that particular form of the central potential. Various forms were proposed for the central potential. Once the Schrödinger equation had been solved for the potential well, and the energy levels obtained, the levels could be systematically filled from the bottom up by the nucleons, taking account of the Pauli Exclusion Principle, in analogy to the filling of atomic shells by electrons. A shell is a collection of energy levels that are close together. The complete filling of a shell would correspond to a nucleus with a magic number of nucleons, just as the inert gases correspond to atoms with filled electron shells. However, the magic numbers arising from this model bore no relationship to the observed magic numbers, no matter what form was chosen for the radial dependence of the central nuclear potential.

The mystery of the erroneous magic numbers, possibly another 'three pipe problem', was solved independently in 1949 by Mayer and Jensen by the introduction of a strong spin-orbit interaction. Nucleons are fermions (see Chap. 10) and possess an intrinsic angular momentum, called 'spin'. The analogy with a spinning top is common but may be misleading. The nucleons also possess an orbital angular momentum which depends on the particular level they occupy in the potential well. In analogy with the Bohr atomic model, this angular momentum might be interpreted as arising from the motion of the nucleon around a prescribed orbit, but the orbital angular momenta are also predicted by the Schrödinger equation. A spin-orbit interaction produces a different energy for the nucleon depending on whether its spin angular momentum and its orbital angular momentum are aligned. When this interaction is taken into account, the nuclear energy levels are displaced so that substantial gaps appear between adjacent energy levels precisely at the observed magic numbers. This was regarded as a great triumph for the nuclear shell theory.

In the 1960s it was predicted by Glenn Seaborg that magic numbers beyond 126 are possible and that 'super-heavy' nuclei might exist with relatively long half-lives. These nuclei would be located in 'islands of stability' on the N versus Z plot (see Fig. 9.4), surrounded by a sea of very unstable nuclei. They might possibly be produced by accelerating heavy ions and colliding them with existing heavy nuclei. One promising possibility is the doubly-magic unibihexium, ^{310}Ubh, with Z = 126 and N = 184, as both 126 and 184 are expected to be magic numbers. The manufacture of elements on this island of stability has been unsuccessful because of the difficulty in obtaining starting nuclei such that when combined together they produce a nucleus with the requisite number of neutrons. A second island of stability is thought to be possible around Z = 164.

The nuclear shell theory can be used to explain other nuclear phenomena besides the magic numbers. For instance, it very successfully predicts the ground state angular momentum of most nuclei.

A situation where some nuclear properties can be predicted by assuming the nucleus is a highly interacting liquid drop, and other properties are explained by nucleons moving independently in a potential well, is clearly not satisfactory. In an attempt to bring the two disparate models closer together, Aage Bohr (son of Niels Bohr, of atomic physics fame), and Ben Mottelson developed a collective model of the nucleus, where the outer nucleons move in a potential well produced by the inner core. In their model the core may be deformed and exhibit collective phenomena, while at the same time the successes of the shell model are preserved.

Examples of nuclear collective phenomena are rotational bands of lines produced by the rotation of a deformed nucleus. These lines are analogous to those seen in the spectroscopy of molecules, so familiar to chemists. Vibrational lines are also observed in nuclear spectroscopy.

We have concentrated in this section on conveying some of the concepts and inherent difficulties that underlie the study of nuclear structure. As we have stressed, the complexity of the physics arises from the highly interacting nature of the nuclear particles, and from their number—too large to enable calculations from basic theory and too small to enable statistical techniques to produce much accuracy.

Ever since Rutherford, experimental physicists have taken great delight in smashing the nuclei of different elements into each other. This has spawned the development of large nuclear accelerators, and of further models to explain the resultant nuclear reactions. These models have their own assumptions and approximations, which we will not discuss here. Suffice it to say that nuclear physics remains a challenging field of basic physical research.

Chapter 10
Fields and Particles

Young man, if I could remember the names of these particles,
I would have been a botanist.

Enrico Fermi

Abstract The development of Atomic and Nuclear Physics has required the assumption of two totally new forces besides electromagnetism and gravity, i.e. the strong and weak interactions. In turn these new forces have required the existence of a few new particles as mediators. These new particles were discovered, but together with them came a "zoo" of other *unrequested* particles. In this chapter we present the "*Standard Model*", which was subsequently developed to classify all these undesired *guests* and describe their nature. The most recent discovery of the Higgs boson is also presented.

10.1 Once Upon a Time

The idea of elementary particles is very old. In the early days of human attempts to understand the physical reality surrounding them, philosophers believed that the apparent enormous variety of substances in nature was in fact the result of different combinations of a limited number of elements. In Greek philosophy there were four elements: earth, water, air and fire. At least, these are the ones listed by Empedocles in the 5th century B.C., although there was some debate on whether all of them were really elementary, or whether they had been generated originally from another unique, more fundamental substance. Aristotle, in the fourth century B.C., introduced one more element: the aether, a constituent of the skies. This fifth element, or "quintessence", has undergone a revival, at least in its name, in the physical cosmology of the last decades, as we shall see in the next chapter.

The idea that the multiplicity of forms in this world comes from a limited number of constituents was not peculiar to the Mediterranean basin. Traces of the Greek elements may be found in the form of the gods of the Mesopotamian cosmologies,

R. Barrett et al., *Physics: The Ultimate Adventure*, Undergraduate Lecture Notes in Physics, DOI 10.1007/978-3-319-31691-8_10

137

one thousand years before Empedocles and his fellow philosophers. The Hindu tradition lists the same Aristotelian five elements in the Vedas (end of the second millennium B.C.). The Chinese also had five elements, even though they were mostly related to changes and transformations of one into another, and the list was different: fire, earth, gold (or metal), water and wood. Classical Japan returned to the Aristotelian quintet.

These are enough examples to show that human culture, since the earliest times, has considered that a limited number of elementary ingredients underpins the multitudinous variety of the world about us.

The Greeks however went one step further. Leucippus and Democritus asserted that everything in nature is made out of *atoms*: tiny indivisible particles (that is the literal meaning of *atomos*), having the properties of the elements. In the last chapter we have seen how in the 19th Century John Dalton resurrected atomic theory, which led to the development of modern chemistry, and to the idea that the combination of the atoms of elements in precise and stable proportions leads to different substances (chemical compounds and their molecules).

As we also saw in Chap. 9, the atom itself was found to have internal structure, and by the 1930s a fairly stable picture of atomic and nuclear structure had emerged: the atom consisted of a central, positively-charged nucleus, surrounded by a "cloud" of electrons. From Chap. 8, we know that the density of the electronic cloud at any point represents the probability of finding an electron at that point. The nucleus was also found to have an internal structure and was constructed from protons and neutrons (or nucleons, as they are known collectively). The masses of nucleons are approximately 2000 times the electronic mass. The scene is completed by a fourth particle, the photon, represented by the quantum of the electromagnetic field, and which had already been discovered at the beginning of the 20th century. The photon is massless, in the sense that it has no rest mass, but of course it carries energy and momentum. In the new (at the time) quantum mechanical framework the photons were the quanta mediating the electromagnetic interaction. So, at that time we had four types of building block from which everything else was constructed.

Theorists had already surmised that there had to be other fundamental particles. Some features of the radioactive β decay suggested the existence of a fourth fundamental interaction, the *weak force*, and one more neutral particle, the *neutrino*, possibly massless. In the framework of quantum mechanics, the strong force had to hold nucleons together by the exchange of massive quanta, which were named *mesons*. At the time when in Europe the Spanish Civil War broke out and the world headed towards disaster, in the field of cosmic radiation a negatively charged, massive object was found, which is now called the μ particle (or muon). At first, the term meson was improperly used for this particle: as we shall see, that name is now used for a category of objects bound together by the strong force, which is not the case for the μ particle.

10.2 The Particle Zoo

After the end of World War II, rather than searching for the most efficient way for us to kill each other, science returned from the battlefield to the field of fundamental physics. A host of experimental evidence led to the discovery of the *pion* (the π meson, mediating the strong interaction) and of the elusive neutrino. The μ particle turned out to be a type of heavy electron (not a meson, i.e. related to the weak, not the strong, force). A bewildering discovery was that, as one went deeper and deeper into the nucleus and closer and closer to the 'elementary' particles (which effectively means going to higher and higher energy interactions), an avalanche of new particles sprang out. They were massive, short lived objects. The increasing mass was consistent with the relativistic identification of energy with mass, i.e. the new particles were not constructed from the broken pieces of the ones forced to interact; they were rather the result of the relativistic conversion of energy into mass. Their instability produced a quick decay of each heavy particle into lighter ones.

Furthermore, when hurling one nucleon against another at high energy in a scattering experiment, protons and neutrons behaved as tiny vesicles, each containing three hard grains. Nobody was (and is) able to isolate any of these "grains", initially called (by Richard Feynman in 1959) *partons* and currently named quarks, but the experimental evidence tells us that they are there. In this way the nucleons lost their quality of "elementary" particles.

We have not mentioned that, since the time of Dirac, a theoretical consequence of quantum mechanics is that each particle has a sort of *specular twin*, endowed with *inverted* physical parameters, i.e. the same mass but opposite charge, spin, etc. Experiment had already confirmed this prediction in the early 1930s with the discovery of the "positive electron" or *positron*. We have discussed the positron briefly in Chap. 9. These "particles in the mirror" are collectively called *antimatter*. Summing up, by the time of the 1960s the original picture of three massive elementary constituents of matter, plus photons, was definitely *passé*, and a plethora of particles was known: μ-ons, π mesons, Λ mesons, Ω mesons, k mesons … This mess of massive, unstable objects was sometimes referred to as the *particle zoo*, and prompted the remark by Fermi at the head of this chapter.

It is now time to leave the historical trail and jump to the present classification scheme for fundamental particles, known as *the standard model* (SM) of particle physics. Just as much earlier Linnaeus had developed a scheme for the classification of all the animals that had been observed at that time, so have physicists developed the SM for particles. Similarly, just as it is impossible to find a totally convincing answer to the question of why some life forms exist and others do not, so the SM does not even try (at least at present) to explain why we have the observed set of particles, and not some other set. It does, however, use tools, such as symmetries, to try to predict the characteristics (parameters) of what might be possible new particles, to tell the experimentalists what to search for.

10.3   The Standard Model

The current understanding of the physics of particles and interactions has produced
a consistent framework within which all known objects fit. The premise to this
classification is that the fundamental interactions of nature are fourfold: gravity,
electromagnetism, strong force, and weak force. Three of these are accounted for in
the SM; gravity remains outside of this model.

Let us start with what is commonly referred to as *matter*. Its primary constituents
are particles. All have a non-zero rest mass, with the possible exception of neutrinos,
which are either massless or extremely light. Furthermore these particles are
endowed with an intrinsic angular momentum (the spin), whose value, in terms of
units of $\hbar$, always turns out to be a positive or negative odd multiple of ½. The value
of the spin is such that all particles obey the Pauli Exclusion Principle (see Chap. 9).
In any system where many identical particles are present, each one of them must be in
a different physical state, i.e. the set of its physical parameters (the *quantum numbers*)
cannot coincide with that of any of the others. Another way of saying the same thing
is to assert that matter particles, also known as *fermions*, obey *Fermi-Dirac statistics*.
In Chap. 5, we saw how the average properties of ensembles of classical particles can
be calculated using statistical techniques. In the case of fermions, the requirement that
two particles cannot occupy the same state, in accordance with the Pauli Exclusion
Principle, results in the need for a different statistical approach, known as *Fermi-
Dirac statistics* for the calculation of their ensemble properties.

This said, matter particles are first subdivided by a grand partition into two
groups: hadrons and leptons. Hadrons include all particles interacting via both the
strong and weak force, whereas leptons are subject only to the weak force. Both
hadrons and leptons interact via the electromagnetic force (if they are electrically
charged, of course) and also gravity, albeit very weakly.

10.4   Leptons

Leptons are indeed elementary in the sense that they do not appear to be made of
simpler components. In the language of the standard model, there are three *generations* of leptons, each made up of two particles. The first generation contains the
electron (e) and the *electronic neutrino* (the symbol is v_e). Then comes the muon
(the μ particle) and the *muonic neutrino* (v_μ). Finally in the third one there is the τ
particle accompanied by its *tauonic neutrino* (v_τ). SM sometimes uses a flowery
language, so one says that the three generations have *different flavours*. The electron, the muon and the tauon carry the same negative charge and the same $\hbar/2$
value for the spin, but have strongly increasing masses. The muon and the tauon are
unstable and after a while they decay into electrons. All six of these particles have
their own antiparticles, even though in the standard theory each neutrino coincides
with its antineutrino and both are massless. The interaction processes, as well as the

Table 10.1 Classification scheme for leptons in the standard model (SM)

Leptons			
		Massive and charged	Massless and neutral
Flavors		e^-	ν_e
		μ^-	ν_μ
		τ^-	ν_τ

decay, are governed by peculiar conservation laws, typical of the electromagnetic and the weak force, and involve neutrinos and antineutrinos.

The masslessness of neutrinos, as already mentioned above, has been called into question by recent experimental evidence. In 2015 Takaaki Kajita of the University of Tokyo and Arthur McDonald of Queen's university, Kingston, Canada were jointly awarded the Nobel Prize in Physics for demonstrating that neutrinos, on the way from the sun to the earth, switch between two flavours (e.g. ν_e to ν_μ). This is only possible if the neutrinos have a mass, however small. This result is in conflict with the SM, which requires neutrinos to be massless. The two experimenters have resolved a puzzle that had existed for decades, the "missing neutrinos" problem. Far fewer neutrinos from the sun were being detected on earth than expected from theory. Now it appears that these missing neutrinos have escaped detection by changing their flavour. The implications for the SM will no doubt be worked through in the coming years.

The whole classification is summarized in the Table 10.1, which holds for leptons.

The same scheme may be used for their antiparticles, which are represented by the same symbols with an overbar and the opposite sign for the charge. The positron for instance is $\bar{e}^+$.

10.5 Hadrons

As mentioned earlier, hadrons are subject to the strong force, as well as the weak and the electromagnetic ones. All hadrons are composites of elementary constituents, i.e. *quarks*[1] (formerly partons).

The quarks are grouped into three generations, divided into flavors, like the leptons. To each generation belongs a pair of quarks, whose names are pure fancy, so in the first generation we find the *up* (*u*) quark and the *down* (*d*) quark. They are stable and electrically charged; however their charge is fractional: either 1/3 or 2/3

---

[1]The name "quark" was proposed by Murray Gell-Mann, one of the originators of the theory, and comes from *Finnegan's Wake* by James Joyce: "Three quarks for Muster Mark". Today most physicists pronounce quark to rhyme with Mark; Gell-Mann preferred the pronunciation "qwork".

Table 10.2 The classification scheme for quarks in the standard model (SM)

Quarks						
	Symbol	Electric charge (units of e)	Spin (units of $\hbar$)	Symbol	Electric charge (units of e)	Spin (units of $\hbar$)
Flavours	u	2/3	1/2	d	−1/3	1/2
	c	2/3	1/2	s	−1/3	1/2
	t	2/3	1/2	b	−1/3	1/2

of the charge of the electron. The second generation consists of a pair of much more massive quarks named *charm* (*c*) and *strange* (*s*); the third generation contains an even more massive pair named *top* (*t*) and *bottom* (*b*).[2] The whole set, with its properties, is shown in Table 10.2.

Of course there are also the corresponding antiparticles: $\bar{u}, \bar{d}; \bar{c}, \bar{s}; \bar{t}, \bar{b}$.

All hadrons produced in scattering experiments and nuclear reactions are combinations in which the quarks appear either in triplets or pairs. To *understand* why they come three or two at a time, but never isolated, it is necessary to go into the theory of the strong force. Here let us simply say that the strong interaction is related to the presence on the quarks of a peculiar quantized "charge", which, unlike the electromagnetic equivalent, is not simply positive or negative, but appears in three kinds. Once again the *strong charge* has been given a fancy name, which is *colour* (but has nothing to do with actual colours); the three types, or colours, are *blue*, *red* and *green*.

An assumed property of the strong interaction is that all directly detectable particles must have a 'neutral colour' which can be obtained by either combining the three different colour charges (just as white may be obtained by superposing blue, red and green), or by combining a colour with its anticolour. Another constraint is that the electric charge of any free particle must be an integer multiple of *e*, disregarding the sign. Finally, the total spin must be an odd multiple of ½ (in units of $\hbar$).

Particles made of three quarks are called *baryons*. An example is the proton, which is obtained by combining two *u* and one *d* quarks, all three with different colour; two of them must have oppositely oriented spin. This situation is represented by the notation: *p:(uud)*, and the charge is correctly +1 and the spin is ½. The other long lived nuclear particle (it is stable within the nucleus, but unstable outside[3]) is the neutron *n:(udd)*, whose electric charge is 0 and its spin is ½.

Particles composed of a quark and an antiquark are *mesons*. For instance, a positive pion is made up of one *u* and one $\bar{d}$, i.e., $\pi^+ : (u\bar{d})$.

---

[2] In the past, the t and b quarks were sometimes referred to as "truth" and "beauty". These names have fallen out of favour. Perhaps they were just *too* ridiculous.

[3] The free neutron decays in about ten minutes via the weak force into a proton, an electron and an electron antineutrino.

The constraints and rules for building hadrons are described by *quantum chromodynamics*,[4] which is the theory of the strong interaction. The corresponding mathematics is not trivial and we shall not go into further details here.

10.6   Bosons

So far we have described particles and mentioned interactions, but in quantum mechanics we know that the interactions themselves are also quantized, i.e. they are transmitted, or mediated, by other kinds of 'particles'. In this group we have already encountered the massless photons, which are responsible for the electromagnetic interaction. The quanta of the three interactions included in the SM carry in general an intrinsic angular momentum, i.e. a spin, but, unlike matter particles, their spin is always an *even* multiple of $\hbar/2$; i.e. they have an integer spin.

This difference in the spin has an important consequence when applying the rules of quantum mechanics. The interaction quanta are not constrained by Pauli's Exclusion Principle; in a given system any number of them may be in the same physical state. In other words, they may have exactly the same set of quantum numbers (i.e., the same values of the independent physical parameters characterizing the state). They are in fact gregarious particles. This feature means that their ensemble properties are not described by Fermi-Dirac statistics (because they are not fermions), nor by classical Maxwell-Boltzman statistics (because they are quanta with *discrete* angular momenta, etc.). A new type of statistics governing large numbers of such quanta has been developed; it is called Bose[5]-Einstein statistics and the particles that obey it are dubbed *bosons*.

Besides our friends the photons, among the bosons we also find the mediators of the weak force, which are the *intermediate vector bosons*: W^+, W^- and Z. Coming to the strong force, the corresponding bosons are called *gluons* from the fact that they 'glue' together the quarks inside the hadrons, and the nucleons within the nucleus. Any hadron (and any nucleus) is then constructed from the appropriate quarks and a cloud of gluons binding them together. The nuclear force binding nucleons together inside the nucleus (see Chap. 9), in the SM becomes a residual effect of the strong force which binds the quarks together into nucleons by the exchange of gluons. Quarks and gluons are mostly confined within the individual nucleons. However their influences can extend beyond the nucleon boundary, resulting in a nucleon-nucleon force. The process is analogous to the force between atoms in chemistry arising from the extension of internal electromagnetic forces beyond the boundaries of the electrically neutral atom.

---

[4]The name is chosen in analogy with *Quantum Electrodynamics*, the highly successful theory of the electromagnetic interaction.

[5]Satyendra Nath Bose was an Indian physicist who made important contributions to the development of quantum mechanics in the 1920s.

Unlike the photons, the other bosons are massive and can carry an electro-magnetic charge, *weak hypercharge* (which includes also the electric charge, but accounts for the weak interaction) and colour charge. The range of an interaction is related to the mass of the corresponding boson. Photons, being massless, lead to the (infinitely) long-range electromagnetic interaction, whereas massive bosons are responsible for short ranged interactions, such as the weak and the strong forces. In a naïve SM one would expect all bosons to be massless, leading to infinite-range strong and weak forces. The mass arises because of interaction with a *Higgs field*. This process is discussed later in this chapter.

10.7   The Role of Symmetries

As we have already mentioned, symmetries play an important role in simplifying and shaping the classification of particles in the SM. A symmetry in general is defined as some operation that, when exerted on a physical system, leaves it unchanged. If we consider a circular cylinder, any rotation about its axis leads to a final configuration identical to the initial one; if we consider an equilateral triangle, any rotation by 120° in its plane about its centre returns the same configuration; if we have a square symmetric matrix, flipping it over about its main diagonal leaves it unchanged; and so on.

In QM symmetries are important both for the wave functions and for the way the different physical operators work. Let us start with the two general categories of particles we have identified: fermions and bosons. The difference, as we have seen, is in the way they group together when many of them are held together in a given system. On the one side we know that fermions have semi-integer spin (odd multiples of $\hbar/2$) whereas bosons have integer spin (even multiples of $\hbar/2$). We have also seen that the idea of 'elementary' particle must be handled with care, so when we discuss fermions and bosons we should remember that they also can be compound objects.

As an example let us consider an atom of He^4. All of its components (i.e., two protons, two neutrons and two electrons) are fermions with individual $\pm\hbar/2$ spin. However, all particles are arranged in pairs (in order to minimize the internal energy) with up and down orientations of the spin. The result is that the total spin in the normal ground state is zero, and consequently the atom turns out to be a boson. If we consider instead the isotope He^3 (which has one less neutron), this pairing of spins is impossible, so that its total spin is $\hbar/2$ and we have a fermion. This difference between the two isotopes is of paramount importance for the properties of the macroscopic Helium fluids.[6]

---

[6]Liquid helium-4, being made up of bosons, exhibits a phenomenon called superfluidity, whereby it can flow up the walls of containers. Liquid helium-3, being made up of fermions which must pair with each other to form a boson, requires a considerably lower temperature to exhibit this phenomenon.

When a number of identical particles are grouped together in a physical system, the whole set is described by its wave function, which is a combination of the wave functions for the individual particles. Recalling the interpretation of the wave function as a probability amplitude, we expect the individual functions to combine according to the rules for the combination of probabilities. If, as in a perfect fluid, we assume that each particle is independent of every other, the probabilities of finding each particle at a given position in a given time interval are independent of each other. The compound probability of having a configuration of the fluid with particle 1 in position 1, particle 2 in position 2, etc. would simply be the product of the individual probabilities. Imagine we throw a die: the probability to get, say, 4 is then p_4[7]; If we throw the die again, the probability to get 2 is p_2, and the probability to obtain the sequence *4,2* is

$$p_{4,2} = p_4 p_2.$$

When we introduced the wave function in Chap. 8, we interpreted it as a *probability amplitude* and its norm squared as a *probability density*. These interpretations do not modify our considerations about the compound probabilities; the probability amplitude for a system of N totally independent particles is given by the product of the individual probability amplitudes:

$$\Psi_{123\ldots N} = \psi_1 \psi_2 \psi_3 \ldots \psi_N \qquad (10.1)$$

The probability density will then be $|\Psi|^2$. So far the order of the indices labelling the particles in the above equation is perfectly arbitrary; i.e., nothing changes if we interchange two particles. In principle, the wave function associated with the whole fluid is a linear combination of wave functions expressing all possible permutations among the individual particles; since all permutations identify the same physical configuration, we obtain[8]

$$\Psi = N! \psi_1 \psi_2 \psi_3 \ldots \psi_N \qquad (10.2)$$

However, we may find situations where the particles are not independent from one another. In particular we have already seen that fermions must obey the Pauli Exclusion Principle; i.e. no fermion in a system can be in the same state as any other fermion of the set. The different possible configurations of such a system are mutually incompatible. In such a case the total probability for obtaining a given

---

[7]If the die is not biased, $p_4 = 1/6$, the same as the probability of having any other result between 1 and 6.

[8]A permutation in a group of N objects represents any possible way the N objects are ordered. The number of permutations among N objects is $N! = N \times (N-1) \times (N-2) \times \cdots \times 1$.

configuration is the *sum* of the probabilities for each possible configuration. Think again of dice. Successive throws give totally independent results, but in a single throw only one of the six possible issues is realized. What is the probability to get a 1 or 2 or 3 in one throw? It will be the sum of the individual, mutually-exclusive probabilities: $p_1 + p_2 + p_3$.

Probabilities, as well as probability densities, are expressed by positive numbers. In quantum mechanics, as we have seen, the probability density corresponds to the square[9] of the wave function $|\psi|^2$. However probability *amplitudes* may also have *negative* values. Any wave, any oscillation, may be partly upward and partly downward. A linear combination of amplitudes can then include both positive and negative terms. For simplicity let us consider a pair of fermions, and let the probability amplitude for the pair be:

$$\Psi_{12} = \psi_1 - \psi_2$$

It is easy to verify that exchanging the two particles reverses the sign of the total amplitude: i.e., $\Psi_{21} = -\Psi_{12}$. The wave function of the pair is *antisymmetric,* or *odd*, for the permutation of the particles. As we shall see, this feature is another expression of the Pauli Exclusion Principle. In fact, if we assume that the two particles are *identical* (which means that all the state parameters are equal) the relationship $\Psi_{21} = \Psi_{12}$ should also be true. This latter constraint requires $\Psi_{12} = \Psi_{21} \equiv 0$, which implies that it is impossible for the two members of the pair to be in the same state. What is true for a pair can be extended to any number of fermions.

The above considerations seem trivial, or even nonsense, whenever the ψ_1's are plane scalar functions: it is always true that $\psi_1\psi_2 = \psi_2\psi_1$, whatever 1 and 2 mean. The situation is, however, more complicated when the wave function must account also for physical properties which are not simply given by plane numerical values or ordinary vectors (as, for instance, the momentum). This is the case when spin is implied.

To better illustrate what we mean here, let us again start from the simple example of dice. It is obvious that, when throwing a die, $p_4p_2 = p_2p_4$ i.e. the fact that you get 4 first does not influence the probability of having 2 next and vice versa: dice throws *commute,* i.e. the order is inconsequential. Suppose now that you have some three-dimensional object, choose three Cartesian axes, then rotate the object by some angle α about the x axis, and then by another angle β about the y axis. It is not difficult to verify that when you perform the operation in the inverse order you end up (excluding special cases) with a different final configuration. Three-dimensional rotations, in general, *do not commute.*

---

[9]For those who already have some knowledge of QM we should specify that the operation to be performed is a bit more sophisticated than a simple square, since the wave function involves also imaginary components. Instead of square we should speak of a *Hermitian conjugation.* Here it suffices to know that the final result is in any case a positive number.

The properties that are accounted for in the state of a quantum object correspond to non-trivial operations performed on the wave function and possess peculiar symmetries.

10.8   Gauge Symmetries and the Fundamental Interactions

We have described a classification of the quantum particles corresponding to the Standard Model in which a relatively small number of ingredients is sufficient to yield a huge number of particles obtained by combining those ingredients. An approach to the above description based on fundamental symmetries is possible. As always in QM we must go beyond our empirical reality into an abstract world of logical properties and operations, but the thread of analogy may guide us.

Let us start from the remark that the formal description of physical phenomena should not be affected by the position of the observer. Locally I need to define a coordinate system to describe what happens around me and this is true also for another scientist located somewhere else. However, we expect the description of reality we are setting up, by our local experiments and theoretical considerations, to be the same in both places. The final conclusions should not be affected by the arbitrariness in the choice of the coordinate systems by different scientists. In General Relativity this is called the *general covariance* of the theory: the equations are the same in spite of the arbitrariness of the choice of the coordinates.

Returning to QM, the simplest situation we may discuss is related to the presence of a wave function ψ, whose information is contained in the *phase*. The phase per se is an angle, whose actual value depends on the arbitrary choice of the initial value, which may vary from one place to another. What we require here is called the *gauge invariance* of the theory. In the simplest example we may think of a unitary circle[10] and an arrow from the centre to a point on the circumference. The theory must be invariant for different orientations of the arrow. The corresponding *gauge symmetry* is technically designated as *U(1)*: *unitary* with 1 parameter (the rotation angle of the arrow). This is the typical symmetry of quantum electrodynamics (and also of classical electromagnetism). There is one coupling parameter (the charge), one gauge field (a four-vector potential) and one uncharged interaction boson (the photon).

A more complicated situation is found when considering the weak force. In general we should remember that most quantum properties reside in an abstract space embedded into the more familiar space-time. In order to describe the weak interaction, theorists have introduced an object, called *isospin*, which may be thought as an arrow pointing from the centre of the sphere to a point on the surface.

---

[10]i.e. With radius 1 in whatever appropriate units.

Again the starting orientation is arbitrary and the physics should be independent of its choice. Now *two* angles are needed to specify the point (the *gauge* parameters), and the corresponding symmetry is called, in the mathematical jargon, SU(2) (special unitary with 2 parameters). Since particles sensing the weak force can also be charged, a unified theory of the *electroweak* interaction (including both the weak and the electromagnetic force) was formulated by Sheldon Glashow, Abdus Salam and Steven Weinberg and brought them the 1979 Nobel prize.

In the new theory, the total symmetry combines the U(1) of electromagnetism with the SU(2) associated with the isospin. An additional coupling parameter is needed to account for the weak interaction and is called (weak) *hypercharge*. Like the electric charge, it corresponds to an additional U(1) symmetry. In symbols the total symmetry is written as $U(1)_{el} \times SU(2) \times U(1)_{weak}$. Here there are *three* quantum interaction vector bosons (in addition to the photon), called W^+, W^-, and Z^0. The W's also carry a positive (respectively negative) electric charge, whereas Z^0 is neutral. The three bosons, unlike the photon, are massive and the presence of the mass is accounted for as a consequence of a *symmetry breaking* due to a peculiar scalar field: the Higgs field with its associated Higgs boson (see next section). The experimental discovery of the W^+, W^-, and Z^0 bosons confirmed the predictions of the theory and brought the 1984 Nobel Prize to Carlo Rubbia and Simon van der Meer.

The reader has undoubtedly realised by now that the technicalities in the foregoing are far from trivial. The essential is that, thinking in terms of symmetries in an internal space associated with each particle, it is possible to reduce the number of *bricks* necessary to build the particles subject to the electroweak interaction to the correct value.

The strong force has been ignored until now, but again the appropriate symmetry in the internal space can reduce the otherwise bewildering zoo of particles to the rather simple scheme presented in Table 10.2. In this case there is no mental image we can rely on to help us picture the symmetry. However the new *landscape* is similar to the one already met for the weak interaction, with one extra dimension. The new symmetry group is called SU(3) and it brings about the six quarks (plus six antiquarks) carrying a threefold "colour" charge, besides the various other "charges" already met. The interaction is mediated by eight 'coloured' vector bosons called *gluons*, which are massive, again as a consequence of a symmetry-breaking governed by Higgs bosons.

Maybe, at this point, the only aspect, which is really clear, is that the real universe has a counterpart in the highly abstract world of symmetries. Hence it becomes spontaneous to ask ourselves why should nature stop at SU(3)? Could SU(5) possibly have also a role at some deeper level? Who knows? After all, the standard model does not include gravity.

10.9   The Problem of the Mass and the Higgs Boson

So far, the basic physical parameters of any particle have been quantized, i.e. they assume discrete values. All charges are multiples of e (or $e/3$ in the case of quarks), and the spin is always a multiple of $\hbar/2$. However, this is not so in the case of the rest mass. The masses of hadrons, as well as of leptons and bosons, are not visibly staggered on some regular scheme based on a given fundamental building block. What does this mean? Should we accept that all the elementary masses in nature are fundamental *'God given'* constants,[11] without any quantization?

This problem was tackled in the 1960s by various physicists, but a solution was specifically proposed in 1964 by Peter Higgs and others. It is not easy, without the help of adequate mathematics, to explain the *Higgs mechanism*, which we have already mentioned in the previous section. We limit ourselves here to stating that the Higgs field has a non-zero value everywhere, which couples to any other existing field. This coupling is the mechanism through which the masses of the quanta of the other fields are defined. Since the Higgs field is itself a quantum field, it can be excited and its excitations appear in the form of bosonic particles. The theory therefore predicts the existence of (at least) one *Higgs boson* whose mass range is also predicted. The Higgs boson became known outside the world of particle physicists when the media took hold of the term *"God particle"*, introduced as a joke in 1993 by Nobel laureate Leon Lederman. Perhaps the nickname can be considered appropriate, since the Higgs field *"gives a mass"* to everything else.[12]

In July 2012, CERN announced the discovery of a new bosonic particle, whose features and behaviour looked very similar to those predicted for the Higgs boson. Since then more experimental evidence has been gathered which supports such an identification. Although the possibility still exists that the newly discovered particle is something other than the Higgs boson, most particle physicists now believe the SM to be substantially complete, the neutrino mass question notwithstanding. Of course, as is always the case with any other apparently well-established theory, we must remain open to the possibility, however remote, that new experiments or observations in the future may require a revision of the currently prevailing SM.

10.10   What About Gravitons?

As we have recalled more than once, gravity is not included in the SM. Nonetheless, in the specialized literature of physics, it is easy to find *gravitons*, i.e. the possible quanta of the gravitational field. *If it existed*, the graviton would be massless (or so it would be expected from standard general relativity) and have spin 2. The point is

---

[11]Fundamental constants of this type are discussed in Chap. 13.

[12]The nickname may bring to the memory of some soccer fans the *mano de Dios* of Diego Armando Maradona in the 1986 match against England.

that there is no evidence, either experimental or even theoretical, of the actual existence of the graviton. The reason why the graviton continues to be mentioned descends partly from an ideological prejudice according to which gravity *must* be quantizable. The other three fundamental interactions are quantizable, so why not gravity? However, the reality remains that nobody has ever succeeded in quantizing gravity. All attempts so far have encountered inconsistencies and infinities. The nature of gravity is in fact different from the other three interactions, as it has to deal with the concepts of space and time, within which the other actors play their role.

10.11   Outstanding Issues

Certainly, the absence of gravity from the Standard Model is a huge problem, but there are also other, more or less disturbing, open questions. A commitment we have imposed on ourselves in this book is not to delve too deeply into mathematical technicalities, so let us consider briefly only a limited number of very general issues.

10.11.1   Antimatter

As we have seen, all fermions are accompanied by their antiparticles. Antimatter is produced experimentally in our accelerators, as well as in high energy processes in the stars or triggered by high energy cosmic rays. However, despite this, the amount of antimatter in the universe seems to be almost negligible. Why? Theorists speak of *spontaneous symmetry breaking* and *charge parity violation*, but no full consensus exists. In any case, would such a theory be an *explanation* or just a consistent and formal *description* of what we actually observe?

10.11.2   Supersymmetry

We have shown a clear-cut distinction between matter particles (fermions) and interaction quanta (bosons). Why should matter particles have only semi-integer values of their spin? What if each of them had a *supersymmetric* partner, with identical quantum numbers except for the spin, whose value would be the closest integer multiple of $\hbar$? Such a correspondence between bosons and fermions would indeed be *Supersymmetry*.

The above hypothesis is not simply based on aesthetics, or an intuitive extension of the concept of symmetry. There are a number of puzzles that could be solved if supersymmetric partners really existed. So far, however, no supersymmetric particles have ever been discovered.

10.11.3 Strings

An alternative to the Standard Model has been developed in the form of the *String Theory*, which includes gravity. The theory, born in the 1960s, has been developed and enriched over many decades. The basic idea is that all particles are quantum states of incredibly small, one-dimensional objects: the *strings*. In order to be consistent, the theory needs space-time to have 11 dimensions, rather than only 4. The reason why seven of the dimensions are not perceived is because they would be *compactified* (i.e. reduced to a very small scale). In a sense, they are rolled up so tightly that at the scale of our experiments they do not show up. The string theory (or more correctly, theories) includes supersymmetry, and has attracted the attention and hard work of a large number of top level theorists. Until now, however, no crucial experiment has been proposed, able to prove or disprove the theory. A weak point of strings is that they "explain" too much, in the sense that an enormous number of possible models exists, in which by adjusting parameters it is possible to account for every positive or negative experiment.[13]

10.11.4 Ptolemy and Quantum Field Theory

From the late antiquity until the end of the Middle Ages, Ptolemy's idea of the skies revolving around the motionless earth grew to a high level of complexity. There were, in the sky, objects (the planets) whose motion violated the simplest view of uniform revolution, in that their actual trajectories presented a number of details not fitting into the initial naïve scheme. Over the centuries, theorists of the time (philosophers) were able to build a theoretically consistent framework explaining almost anything that could be seen in the sky with the available instruments (direct observation, astrolabes, armillary spheres, sextants, etc.). The theory, as we will see in the next chapter, included a complete machinery accounting for all details of the apparent motion of the celestial bodies. As we know, the Copernican revolution (anticipated eighteen centuries earlier by Aristarchus) gave a much simpler explanation.

It is difficult to escape the impression that, in spite of the enormous progress and success enjoyed in the last century, the present state of theoretical particle and field physics is living through a type of *Ptolemaic phase*. Perhaps in the future another "Copernican revolution" will come along to shake us all up?

---

[13]One may recall Wolfgang Pauli's comment about "untestable" theories in Chap. 3.

Chapter 11
Cosmology

My view is that if your philosophy is not unsettled daily then you are blind to all the universe has to offer.

Neil deGrasse Tyson

Abstract After having explained the principles of relativity and of quantum mechanics in Chaps. 7 and 8, the known physics is applied to the description of the visible universe. The progress of the knowledge gained from observation is followed through the course of the 20th century showing the triumph of the theoretical insight of general relativity, but also the appearance of new unexpected features visible on the highest scale. The role of dark matter and dark energy is discussed and the generalization of the cosmic models to include them is presented. The difficulties associated with the mismatch of quantum physics and general relativity showing up in the early stages of the universe are considered.

11.1 The Time of Myths: Cosmogonies and Cosmologies

Human beings have been wondering about the world in which they live since prehistoric times. Even when life was short and threatened brutally by all kinds of impending dangers, they engaged their thoughts not only in the quest for better weapons, housing and foraging, but also in asking questions about the meaning and nature of what they could observe: the sky with all its stars, sun and moon, the variable and at times pitiless weather, the sea of apparently boundless water. Since they were alive, they interpreted everything around them as also being animated. Unable to understand natural bodies and their behaviour, they attributed to them a divine nature. Their description of the world was mythological, i.e. intertwined with symbols and legends. To answer the question of its origin, some primordial form of cosmogony[1] was assumed: the shapeless chaos of Greek mythology, the waters of the Bible, the *qi* of the Chinese tradition. Gods were also born, as was any other form of life, as the result of some inexplicable and sudden event.

---

[1] Cosmogony: the branch of science that deals with the origin of the universe, especially the solar system—O.E.D.

© Springer International Publishing Switzerland 2016
R. Barrett et al., *Physics: The Ultimate Adventure*, Undergraduate Lecture Notes in Physics, DOI 10.1007/978-3-319-31691-8_11

Ancient mythology is a fascinating subject, but what is interesting for us here are just a few intuitions, wrapped in the veils of symbolism, which later inspired philosophers and finally even modern science. The primordial Greek divinity *Chronos*, who devours his children, is *time*. Could we say that the myth hints at the arrow of time and perhaps even to the second principle of thermodynamics that we encountered in Chap. 5? The latter can perhaps be glimpsed also in Hesiod's five ages of the world, going from the best (golden) to the worst (iron). Something similar could be argued for the four epochs of the Indian Vedas and Puranas.

Both the Chinese and the Hindu traditions conceive the evolution of the universe as a cyclic process. The Hindu cosmology even quantifies the duration of each cycle in terms of *billions* of years. At the end of each cycle the universe is destroyed by fire, and then after an interlude, it is created again. Both the time-scale and the cyclic nature recall one of the Friedman-Le Maître-Robertson-Walker solutions to the Einstein equations.[2]

We cannot seriously attach much consequence to these correspondences, but it is instructive to realize that human beings have long been awestruck by the mysteries of nature, even if their confused and emotional approach was very remote from our present formal mind-set. Maybe one could even argue that it is this yearning for knowledge that defines the species: *homo sapiens*.

11.2   Ancient Rational Cosmology

The first attempts at a rational description of the universe may be found among the Greek philosophers of a few centuries BC. Observations of the sky suggested that the earth had to be a sphere and that the skies revolved around it diurnally. The plural (skies) is appropriate, since the most conspicuous celestial bodies (the sun and the moon) had additional intrinsic cyclical rotations, which meant that each of them was attached to a different layer (sky), revolving separately. However the observations also revealed the presence of a small number (five) of *wandering stars, or planets.*[3]

The movements of the planets were not easy to understand, since sometimes they even went in the *wrong direction* with respect to the other stars. As early as the third century BC, the Greek astronomer Aristarchus of Samos surmised that their apparently queer behaviour could be explained by assuming that the planets revolve around the sun, rather than the earth. However the prevailing view, rooted in Aristotle's teachings, was against such an assumption and kept the earth at the centre of the universe. At the beginning of the 2nd century AD, Claudius Ptolemy, an astronomer of Alexandria, elaborated a geocentric description of the universe, in which an articulated machinery (to be enriched in the ensuing centuries), made up

---

[2]These are discussed later in this chapter.

[3]From the Greek term *planētes asteres* meaning "wandering stars".

of *epicycles*, *equants*, *deferents* (or *excentrics*), explained all of the apparent anomalies in the motions of the celestial bodies. By the onset of the Middle Ages, Aristarchus had been completely forgotten and Ptolemy's was the standard philosophical description of the universe, strengthened by an alliance with theology.

It was only in the 16th century that Copernicus revived Aristarchus's heliocentric model. The troubles accompanying the Reformation and conflict between different political powers (including the Roman Church) made the *Copernican Revolution* particularly dramatic, as we know from the history of Galileo. At the end of the 16th Century, Giordano Bruno, who was charged with heresy and burned at the stake in AD 1600, asserted that the universe had to be infinite and contained an infinite number of worlds (suns and earths). He came to this conclusion not by means of astronomical evidence, but through philosophical arguments.

From the 17th century, modern science became pervasive in all areas, including astronomy and cosmology. Telescopes allowed a much more powerful observation of the sky. Newton's universal gravitation shed new light on celestial mechanics and, at the end of the 19th century, the description and interpretation of the universe in scientific terms began.

11.3   Is the Universe Infinite, Homogeneous and Eternal?

The common representation of the Universe among astronomers (and scientists in general) during the 19th century was that of an infinite, substantially homogeneous expanse of stars similar to the sun, with no location within the universe being special, or more *privileged*, than any other. Every object within the universe had its own evolutionary path, but anything dying was always replaced by something else being born, so that on the average the Universe would always appear to be the same, lasting forever, or at least indefinitely.

These convictions were only marginally based on observations, but relied strongly on implicit or explicit a priori tenets, moulded by the previous history of emancipation of modern science. On one hand, astronomers had verified that most visible stars displayed zero parallax during the year, from which they deduced that these stars lie at enormous distances from the earth (and the sun). On the other hand, there was no logical reason for any *privileged* place, or "centre", of the universe. Similarly, why should the presence of stars end somewhere, leaving an infinite emptiness beyond? The universe must have no border and no centre.

What held in space should hold also in time. There was no *special* moment here either; the universe had to be in a stationary state. In this case, the argument was more delicate and sensitive. If not to reject, at least to mark a complete separation from religion, no *origin* could be envisaged by science.

This image of the universe somehow paralleled the one scientists had settled on for the earth. From the biblical age of 6000 years or so, the age of the earth had been pushed back by the newly-born geology of Charles Lyell towards an undefined remote past. The principle of *uniformitarianism*[4] contended that the crust of the earth had been transformed in the past by the same processes active today, amounting to a type of stationary state (or at least an indefinite age) for our planet.

Yet the picture was not fully consistent. A glance at the night sky revealed an appearance that was not really homogeneous. For instance, there was that whitish stripe crossing from one section of the horizon to the other (our Milky Way). Ancient Greeks had called it γαλαξίας κύκλος, which literally means *milky circle*, and the myth recounted the story of a drop of milk fallen from Juno's breast while she was nursing Hercules. Other fainter clouds were also visible in the sky: the big one in Andromeda, and smaller ones such as the Magellanic Clouds[5] in the southern hemisphere. Once again it was a Greek philosopher, Democritus who, between the 5th and 4th centuries BC, had suggested that the Milky Way could be made of innumerable distant stars concentrated in a wide band of the sky. Modern observations and telescopes confirmed Democritus's intuition.

The so-called *Olbers paradox*,[6] which can be summarized in the apparently silly question: *Why is the sky dark at night?* also cast doubts on the conjecture of an infinite and uniform distribution of stars in the universe. The answer is by no means trivial since, if you assume that there are infinitely many stars scattered everywhere and at all distances, all lines of sight from your eye end sooner or later on the surface of a star. Of course, the farther you go, the smaller the apparent size of the star. However, even if the apparent size is reduced to an infinitesimal point, the luminosity will always be that of the surface of a star. As a result the sky should always be as bright as the surface of the sun.

A possible objection might be: "well, you are assuming that the skies are perfectly empty and transparent. There could be (and in fact there is) opaque dust scattered around. Its density is quite small, but still sufficient to screen off distant stars". This objection is correct, but it stumbles against the alleged eternity of the universe: if its present age is infinite, dust, radiation, stars, planets, everything, should be in thermal equilibrium. A dust grain absorbs radiation having the same temperature as the stars, warms up a bit, then re-emits the absorbed energy in the infrared (even far infrared) band of frequency (not visible to our eyes). However, if the system is in thermal equilibrium, dust will have reached the same temperature as

---

[4]Uniformitarianism is the scientific observation that the same natural laws and processes that operate in the universe now have always operated in the universe in the past and apply everywhere in the universe. (Wikipedia Definition.)

[5]The name comes from the fact that Ferdinand Magellan noticed the "clouds" during his crossing of the Pacific Ocean (1520–1521). They appeared as little clouds but, night after night, the place they occupied in the sky was always the same. The Persian astronomer Al Sufi had however already seen these "clouds" in the 11th century. The clouds have featured in the mythologies of the Australian Aboriginal peoples, probably dating back 40,000 years.

[6]Actually, this paradox, in one form or another, had been considered since at least a couple of centuries before Olbers, who formulated it in 1823.

the incoming radiation, and will emit at the same frequency as is incident upon it. Again the sky should be as bright as the surface of a star.

Another possible objection is: "light takes time to travel from the source to our eyes, so the photons from the farthest stars have simply not yet arrived on earth". This remark implicitly assumes an origin of time, i.e. a beginning, thereby denying the eternity of the universe in the past.

Another open problem arose in the 19th century from classical thermodynamics (see Chap. 5). Clausius's formulation of thermodynamics states that heat in an isolated system always flows from warmer to cooler bodies. In practice, within an isolated system—the universe is isolated by definition—all processes cannot be globally reversible: there is an arrow of time. Starting from any given situation with a number (possibly infinite) of hot stars and a number of other cold bodies (planets, asteroids, dust, etc....), as time goes on the former get cooler and cooler, and the latter warmer and warmer. Asymptotically the universe reaches a *thermal death* state, in which everything has the same (low) temperature and nothing happens any more.

This argument tells us that the universe, at least for what concerns the variety of its forms, will have an end. But logic requires that if something is due to end at a given time in the future, it must have had a beginning at a given time in the past. Both the beginning and the end may have an asymptotic nature: a cold *dead* universe without inhomogeneities and motion may remain *frozen* forever, but going back in time, it would become hotter and hotter with an ever growing temperature difference between the hottest and coldest spots. What matters is that the past is dramatically different from the future, in contrast with the idea of a stationary state.

All of the above objections did not really trouble 19th century scientists and in the early times of Einstein, the stationary, uniform and infinite universe was the common wisdom, shared by Einstein himself.

11.4 Relativistic Cosmology

We have already summarized Einstein's General Relativity in Chap. 7. We recall here that a fundamental novelty of his theory was the "geometrisation" of gravity. He succeeded in bridling time, reducing it to one of the four dimensions of space-time. Leaving aside for simplicity two of the spatial dimensions, space-time can be visualised as a "surface" curved by the presence of matter and energy. The curvature is the gravitational field. The reciprocal interaction between space-time and matter/energy is described by Einstein's equations, which may be solved analytically to find the gravitational field of a spherical non-rotating, or axially-symmetric rotating, mass, or even of the grand-scale universe.

Locally the distribution of matter is highly uneven and anisotropic, but for ever increasing volumes the average density tends to become uniform and constant. So, if the scale is large enough, the matter content of the universe can be modelled as a homogeneous dust made of stars with negligible mutual interaction. Under this assumption Einstein's equations yield a *simple* analytical solution. The analysis was carried out independently by the Russian physicist and mathematician Alexander Friedmann, and by the Belgian priest and astronomer Georges Lemaître.[7] Later, Howard Robertson and Arthur Walker gave a complete formal justification of the solution, which is now known as the *Friedmann-Lemaître-Robertson-Walker* (FLRW) model.

The relevant aspect of a universe that is uniformly and isotropically filled with massive dust is that it must either expand or contract; no steady state is allowed. Pictorially, we can think of it as an inflating (or deflating) balloon. The surface represents space (which is actually three-dimensional), with time (the cosmic time) along the radial direction. The dust grains, locally at rest on the surface, drift apart (or get closer to one another) as the expansion (or contraction) proceeds. As we shall see, this representation corresponds to just one out of three possible solutions, but tells us the essential: no steady state solution is allowed. Although this result was recognized to be mathematically correct by Albert Einstein, he did not like it because in an expanding universe, going back in time, the matter/energy density becomes infinite at a finite point in the past. Such an infinite density was considered unphysical.

Einstein's preferred universe was a "hypersphere"—a structure similar to a sphere but with a three-dimensional "surface". Space had a *finite* extension, although *unbounded* (similar to the surface of a sphere), and the size of the universe was constant in time. In such a *closed* universe, a light ray sent in a given direction (and not absorbed or scattered by any obstacle anywhere) would always come back to the starting point. The radius of the closed trajectory would be inversely pro- portional to the square root of the average matter density. However, the weak point of such a gravitationally bound universe is that any small positive fluctuation of the matter density produces a positive feedback, leading to an indefinitely growing density and a global collapse. In order to avoid this, Einstein introduced an addi- tional term into his equations, known as the *cosmological constant* Λ. For a solution of the equations to exist, this new constant had to be *fine tuned*, i.e. it had to have exactly the value:

$$\Lambda = 4\pi \frac{G}{c^2} \rho,$$

where ρ is the average matter density of the universe.

The new term is perfectly compatible with General Relativity and mathemati- cally consistent, but its physical interpretation is rather strange. It looks as if a perfectly homogeneous fluid permeates all space-time, with positive energy density

---

[7]The solution by Friedmann is dated 1922 (published in 1924); the one by Lemaître was published in 1927.

(which is OK), but *negative* pressure, i.e. it introduces a repulsion everywhere that exactly counterbalances the gravitational attraction among the grains of the universal dust.

What is worse, however, is that Einstein's solution is unstable; any arbitrarily small perturbation of the matter density, or of the value of the cosmological constant, produces an irreversible switch to an expanding or contracting universe. The static and closed universe solution was therefore abandoned in favour of Friedmann and Lemaître's solutions. Einstein, according to a later testimony by George Gamow, called the cosmological constant the biggest blunder of his life. Today, however, Λ has been *revitalised*, as we shall see in a later section.

11.5   The Universe Expands, but Are We Really Sure?

Huge advances in the design of telescopes and in the observational techniques during the first decades of the 20th century brought tremendous progress in the study of the sky. First, the number of visible objects formerly known as *nebulae* steadily increased, and it became clear that they were comprised of stars. Next, an ongoing ancient debate on their whereabouts came to an end with the realization that they are located much farther away from us than are our surrounding stars. In fact, we now know that stars are grouped into huge "island universes" (as Immanuel Kant had called them) or *galaxies*, as we call them today from the Greek name for the Milky Way. The Milky Way, visible in our sky, is simply the inner part of our galaxy, seen from within.

Galaxies are enormous aggregates of gravitationally bound stars. They are grouped according to a morphological classification based on their shape, but in general they resemble flat disks, with or without a central bulge and/or spiral arms. Astronomers believe that our galaxy, seen from outside, would look more or less like the galaxy NGC 6744, shown in Fig. 11.1. The Milky Way is a *barred spiral galaxy*. Its diameter is in the order of 100,000 light years and our sun is located in the inner side of one of the arms of the spiral at some 27,000 light years from the centre.

Stars in a typical galaxy, such as the Milky Way, are counted in hundreds of billions, and within our range of sight in the universe, there are approximately a hundred billion galaxies.[8] Figure 11.2 offers a picture taken during the ultra-deep research program of the Hubble space telescope, showing distant galaxies visible in the near infrared domain. The portion of sky visible in the picture is approximately one tenth of the diameter of the full moon. Only a handful of the visible objects are stars of our own galaxy; the rest, even the faintest little spots, are far away galaxies.

The existence of galaxies does not introduce much change to the FLRW model; simply, instead of considering a dust of stars, the grains of the "universal dust"

---

[8]It is the same order of magnitude as the number of neurons in the human brain.

Fig. 11.1 The NGC 6744 galaxy photographed by the Wide Field Imager on board the Hubble Space Telescope (The figure is contained in the file "Wide Field Imager view of a Milky Way look-alike NGC 6744.jpg" available on Wikimedia Commons)

become whole galaxies. Another important discovery by the American astronomer Edwin Hubble had a far stronger impact on the debate concerning Relativistic Cosmology. Hubble was already well known for his fundamental contributions to the very difficult task of assessing galactic distances.[9] In fact he had played an important role in demonstrating that galaxies are far more distant from us than the stars of the Milky Way.

In 1929 Hubble published the results of countless new observations, made together with his assistant, Milton Humason, on the light received from galaxies. Light from very bright objects, such as stars, can be analysed spectroscopically (see Chap. 9) and found to contain the *signature* of the emitting chemical elements. In addition, since light is an electromagnetic wave, it also displays another phenomenon typical of wave propagation: namely, the Doppler shift, which we have also encountered in Chap. 3. This effect occurs when we have a moving source of radiation. If the source is receding from us, the received frequency decreases, and if

---

[9]See Chap. 3.

Fig. 11.2 Ultra-deep field picture of the sky in the near infrared domain, taken by the Hubble space telescope. Apart from a few stars (displaying a diffraction pattern) of our own galaxy, what we see are all galaxies (The image is downloadable at http://spacetelescope.org/images/heic0406b/)

the source is approaching us the frequency increases. In the case of visible light, the former situation implies that all colours shift towards the red; the latter yields a shift towards the blue.

Observing the sky in the scenario of a static universe, we would expect to find approximately equal numbers of stars approaching us as receding from us. Studying the spectrum of the received light from the stars, it should be possible to recognize the *signature* of hydrogen (and of other elements), but we would expect the light to appear redder or bluer than in a corresponding terrestrial laboratory experiment. The magnitude of the frequency shift allows us to calculate the radial velocity of the source. Carrying out these measurements with galaxies, Hubble found that in almost all cases, the light was red-shifted, which meant that practically all of the galaxies were receding away from us. He also noticed that the larger the distance, the greater was the red-shift. By this means he could establish a simple proportionality relation between the recession velocity V_R and the distance D:

$$V_R = H_0 D$$

where H_0 is known as the *Hubble constant*.

Hubble's law had been anticipated a couple of years earlier by Georges Lemaître, while analysing the implications of his expanding universe. The red-shift of galaxies became a turning point in the history of Cosmology. Einstein abandoned his cosmological constant and the Friedmann-Lemaître solution of the field equations of General Relativity lost its status as a mathematical curiosity and became accepted fully-fledged into the domain of physics. The expansion of cosmic space became a *fait accompli*. In our earlier balloon analogy, the behaviour of stars corresponding to Hubble's law is seen in the increasing separation of specks scattered on the surface of an inflating balloon.

An immediate consequence of the *expansion of space* was the *finite age* of the universe. The current estimate of this age is 13.8 billion years. Going backwards in time, the origin of the universe corresponds to an infinite density condition with the mass/energy of all galaxies squeezed into a dimensionless point,[10] which we might call a *space-time singularity*. Of course, physicists did not (and still do not) feel at ease with the idea of an origin of time, since it invokes scenarios outside the domain of science.

Upon hearing that the universe has an origin, the first *naïve* question that comes to mind is "what happened *before* the origin?" Such a question does not make much sense and is contradictory. The word "before" implies time, but the singularity is also the origin of time. In a relativistic approach the origin, in spite of being a singularity, is not as upsetting as it appears at first sight. Recalling that General Relativity is geometry in four dimensions, the situation may be grasped from the simple bi-dimensional example shown in Fig. 11.3 (a type of Chinese hat with a cusp). The cusp is indeed a geometrical singularity, but nobody is upset by its presence. All lines stemming out of the cusp and extending along the maximal slope may be used as lines along which the distance on the surface from the cusp can be measured. Of course, the image is in two dimensions and Euclidean. However the analogy (with some *caveats*) holds also for space-time in four dimensions. Cosmic time is measured along the lines originating in the singularity. The whole four-dimensional manifold is there, including time, and there is no global *before* or *after*; one simply has one specific shape rather than another.

---

[10]The image of a physical point concentrating all the matter of the universe is the most intuitive, but it is just a special case. The topology of the initial singularity may not be that simple, so that it is more correct to speak in terms of infinite density configuration.

Fig. 11.3 A two-dimensional view of an indefinitely extended surface with a cusp, as a model representation of the initial singularity

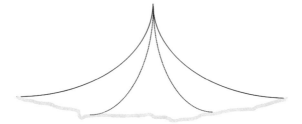

Although the red shift has now become a routine tool in Astronomy, not everybody accepted its interpretation in terms of the expansion of the universe, and several scientists proposed alternative explanations. An example is the "tired light" hypothesis put forth by Fritz Zwicky in 1929.[11] He remarked that the reddening of light from distant galaxies could be due to loss of energy by the photons, through various mechanisms acting along the travel path, such as scattering against dust particles and photon-photon interaction. A drawback of the scattering mechanism (and indeed of any other, unless new kinds of interactions are assumed) is that the image of distant light sources should be blurred with distance, which is not the case. The "tired light" explanation for the cosmic red-shift was progressively marginalised and finally abandoned.[12]

Another attempt at preserving a static universe was made by the British scientist Fred Hoyle. In 1948 he proposed a mechanism that would have guaranteed, in his opinion, a *steady state* for the universe. He accepted that the cosmic red-shift was due to an expansion, but rejected the corollary of a hotter past and cooler future. In particular, he rejected the idea of an initial singularity, which he jeeringly nicknamed "big bang", never suspecting the term would soon become a hit, even among the supporters of the FLRW model. Hoyle maintained that a steady state universe could be compatible with the drifting apart of galaxies if, due to a "creation field", *new* matter were spontaneously and continuously generated in the intergalactic space.[13] The amount of newly created matter required to ensure the steady state condition was very little: approximately one hydrogen atom per cubic kilometre per year. This is so small as to be totally unobservable. The steady state theory was thus based on the assumption of an unobservable phenomenon, and as such might be thought incompatible[14] with what we call *Physics*. However the theory does predict

---

[11]The expression "tired light" was actually introduced by Richard Tolman in the early 1930s.

[12]In truth, in cosmology and theoretical physics, nothing is ever really "abandoned". A relevant example we have already mentioned is the cosmological constant, which has now become part of the so called concordance model or standard cosmological model. (See later in this chapter.)

[13]Independently, similar ideas were also considered by Hermann Bondi and Thomas Gold.

[14]We have discussed this incompatibility issue in Chap. 3. Atoms and neutrinos, at the time of their initial introduction into physics were undetectable. They did however have consequences which were readily detectable, and became accepted by physicists long before they were detected in their own right.

observable differences with the FLRW model. Newly discovered radio sources (quasars and radio galaxies) were associated in the big bang theory with the early stages of the universe; they were therefore expected to be found only at large distances. From the Steady State Theory, these strange objects were expected to be uniformly distributed throughout the universe.[15]

The steady state theory was finally swept away by new observational evidence emerging in 1964 and analysed and perfected afterwards: namely, the *relic radiation* or *cosmic microwave background* (CMB). Hoyle's reluctance to accept the demise of his theory shows that even a front rank astrophysicist like Fred Hoyle may be not immune from prejudices.[16]

11.6   The Cosmic Microwave Background

Let us begin our story in the so-called *recombination era*, i.e. 380,000 years after the Big Bang (or 13.8 billion years ago). At that time, according to the FLRW model, matter was in the form of a rarefied plasma and the estimated temperature was around 3000 K. Above that temperature, i.e. before the recombination era, charged particles (mostly electrons and protons) could not bind into atoms because the thermal radiation they continuously emitted and absorbed had enough energy to prevent the formation of stable bonds. The cosmic medium at that time resembled a bright white fog. Afterwards the temperature decreased below the threshold of 3000 K and the average energy per photon was no longer large enough to break the bonds when they formed, so that hydrogen atoms started to appear. Also photons, not having enough energy to ionise hydrogen, could no longer be absorbed, but only scatter elastically and the universe started to become transparent.

Light escaping from the recombination era and arriving to us without being absorbed undergoes a red shift (similar to that of light produced by stars in distant galaxies). However, in this case the distance is so great that the radiation wavelength is stretched by a factor of approximately 1100; what initially was visible light now appears as microwave radiation. The equilibrium radiation field of this primordial hot plasma displayed a typical thermal radiation spectrum, in which the peak of the Planck distribution of frequencies depends only on the average temperature. The red shift reduces the initial 3000 K radiation spectrum to one corresponding to a temperature of just $T_{CMB} = 2.72548$ K today.

---

[15]Steven Weinberg wrote in 1972: "In a sense, this disagreement is a credit to the model; alone among all cosmologies, the steady-state model makes such definite predictions that it can be disproved even with the limited observational evidence at our disposal.".

[16]The Steady State Theory received a new incarnation in the Quasi-steady state cosmology (QSS) proposed in 1993 by Fred Hoyle, Geoffrey Burbridge, and Jayant V. Narlikar. It was intended to explain additional features unaccounted for in the initial proposal, but ran into further difficulties and is not generally accepted.

In 1964 Arno Penzias and Robert Wilson, while testing a microwave antenna at the Bell Telephone Laboratories, accidentally found a "noise" coming uniformly from every direction in the sky. That "noise" was quickly recognized as being the cosmic microwave background (CMB), a.k.a. the *relic radiation*, arriving to us after having survived more than thirteen billion years of travel. Penzias and Wilson received the Nobel Prize for their discovery in 1978. Their experimental observation sounded the death knell for Hoyle's steady state universe.

Also, Olber's paradox is explained at last, thanks to the expansion of the universe and the ensuing red shift of light. In a clear night, the sky appears totally black because our eyes sense only a very small portion of the electromagnetic spectrum. For a microwave antenna, the sky is uniformly bright, not as the surface of a suitably red-shifted star, but as a red-shifted hot plasma transitioning to a neutral gas.

The uniformity of the CMB both in intensity and temperature is amazing, with fluctuations of only 1 part in 100,000. This uniformity raises an interesting question. As we have said, the recombination era and the corresponding *last scattering surface* (another term used to designate the onset of the transparency of the universe) are located 380,000 years after the Big Bang. Distant portions of the sky, from which the radiation comes, were at that time separated by more than 380,000 light years. Since the Big Bang, no causal interconnection can have taken place between the regions, since no physical "messenger" can travel faster than light. Then how is it possible that those independent regions can have kept such an amazing synchrony, arriving at the same average temperature in the same time? We shall come back to this point in the next section.

Yet differences among various areas of the sky do exist. Evidence of this inhomogeneity can be seen in Fig. 11.4, where the temperature of the radiation coming from the whole cup of the sky is shown. To be sure, the unprocessed data from the sky are not as clean as in Fig. 11.4, since there are also microwave emitters in the foreground, primarily from the Milky Way. In order to study the primordial radiation field, these nearby sources had to be removed from the data. The picture, taken from the NASA WMAP survey, presents the temperature of the radiation using false colours: the red areas are the warmest, the dark blue the coolest. Remember that the difference between the coolest and the warmest is less than 1 part in 100,000, so the differences have been enormously magnified in order to make them perceptible to the eye. To understand the figure, imagine being at the centre of the celestial sphere, cut it from one pole to the other, then open and flatten it. In this way, the right border corresponds to the left one.

Despite their smallness, the anisotropies in the CMB carry a wealth of important information about the primordial universe. An immediate remark, looking at Fig. 11.4, is that the distribution of warm and cool areas is not really random: some irregular structures seem to exist. First, there is a concentration of red and yellow blots on the right side of the image and an area of prevailing dark blue in the middle (and again on the right border). Returning to the initial three-dimensional sphere, the warmest and coolest areas turn out to be in opposite directions in the sky: this is the *cosmic anisotropy* of the CMB.

Fig. 11.4 Distribution of the equivalent temperature of the CMB across the sky. False colours have been used: *red* means warmer; *blue* means cooler. The measurement has been made by the NASA space-based microwave telescope WMAP ("Ilc 9 yr moll4096" by NASA/WMAP Science Team—http://map.gsfc.nasa.gov/media/121238/ilc_9yr_moll4096.png Licensed under Public Domain via Wikimedia Commons—http://commons.wikimedia.org/wiki/File:Ilc_9yr_moll4096.png#/media/File:Ilc_9yr_moll4096.png)

The explanation is simple. The thermal radiation bath, in which everything in the universe is immersed, introduces a privileged reference frame. Relativity is not violated, since the local speed of light is the same in all directions for any observer, but it is not so for subluminal (i.e. $V < c$) objects. Their motion with respect to the background can be detected. The earth is indeed moving with respect to the microwave background, so the incoming radiation is Doppler red-shifted in the direction of motion and blue-shifted in the opposite direction. This Doppler frequency shift is superposed on the cosmic red-shift. The "absolute" motion of the earth (i.e. the motion with respect to the background) corresponds to a velocity of 371 km/s in the direction of the constellation Leo. Of course, the earth is moving around the sun and together with it; the sun is in turn moving within and together with our galaxy. These motions are local and superimposed on the *cosmic flow*, which is due only to the expansion of the universe.

Once we know its origin, we may as well subtract this Doppler anisotropy, so that we are left only with the primordial anisotropies. The study of the latter is extremely important in many aspects, which we cannot discuss here. However, a quick qualitative remark is that the tiny density fluctuations present at the recombination era are the "seeds" of the future galaxies. In fact, any density excess, via a gravitational positive feedback, would tend to grow, attracting matter from surrounding less dense regions and in the process give rise to large scale *structure*.

Also, the pattern of density fluctuations of the last scattering surface carries the imprint of what happened *before* that era. Nothing else can arrive to us from earlier times closer to the Big Bang. The last scattering surface acts as a bright impenetrable curtain that prevents us from plunging deeper towards the singularity. The history of the universe before the recombination may be inferred to some extent from the accepted laws of physics, but cannot be observed.[17]

11.7   The Standard Universe Before 1998

We can now assemble facts and theories to construct a consistent cosmology as was accepted at the end of the 20th century. General relativity is the basic conceptual framework; the expanding universe is well established in the FLRW formalism. A consistent chain of physical events may be described using known physics (including quantum mechanics for nuclear and sub-nuclear forces) from very close to the initial singularity up to the present time. Not everything is clear and understood, but the fundamental pillars are well established.

The initial singularity remains outside the domain of physics, since nobody knows what to do with an infinite energy density. Immediately following the singularity comes an incredibly short time, the so-called *Planck era*, lasting 5.4×10^{-44} s, where we do not know how to apply the laws of physics.

Today the gravitational interaction is macroscopic and universal, whereas the other fundamental forces dominate at short distances. If we compare the gravitational attraction between two protons with their mutual electric repulsion, we find a ratio of $1:10^{36}$. The imbalance is even stronger in the case of the strong nuclear force. The emergence of gravity as a potent force in the everyday world is because the other forces are rapidly screened by the interplay of complementary charges, while gravity, whose nature is geometrical, is not.

At the molecular, atomic and subatomic scale, gravity is totally negligible. However, if we imagine squeezing matter to the enormous densities expected close to the singularity, the gravitational energy density (or, in geometrical terms, the curvature in the proximity of the space-time cusp) becomes comparable and even greater than the energy density of the strong, weak and electromagnetic interactions together. We do not know how to treat such a state because we should consider the interaction among the four forces, but three of them are described by QM, while the other, gravity, has so far stubbornly resisted any attempt to bring it into the framework of quantum physics.

Following the Planck era comes a time interval, lasting approximately from 10^{-36} s to 10^{-32} s after the Big Bang, in which the physical scenario still remains

---

[17]This is certainly true for the information carried by electromagnetic signals, since photons are stable quanta of a long range interaction. Nothing similar is possible for the strong and weak interactions, which are short range interactions and whose bosons are generally short living objects. Various non-standard theories predict the existence of other "messengers" from the very early times of the universe. Besides these it has been conjectured that gravitational waves could also exist as a relic of the pre-recombination era. However at present no observational evidence exists for any of this.

rather uncertain, but where theorists have discovered a mechanism capable of solving the puzzle of the homogeneity of the CMB. In that interval, the theory asserts that the universe should have undergone an exponential expansion, increasing the scale of spatial distances by a factor of at least 10^{26}. Such an enormous expansion rate, translated into the recession velocities of the components of the universe, gives values well above the speed of light. Is that in contradiction with relativity? No. The theory of relativity tells us that no object *in* space-time can travel faster than light. In the case of the expansion, however, it is space itself that expands, and so no logical a priori barrier exists for the expansion rate. This type of tremendous expansion was called *inflation* in the theory initially proposed by Allan Guth and Andrei Linde in the 1980s.

In practice inflation smooths out the density fluctuations present in the cosmic fluid at the beginning of the exponential expansion, leading at a cosmic scale to the homogeneity we observe in the CMB. At the same time, the tiny residual inhomogeneities are frozen into the universe, in the sense that regions carrying them are brought out of causal connection with one another for a long time into the future (as nothing can travel faster than light *in* space-time).

According to Guth and Linde's theory, inflation is the result of a peculiar field called the *inflaton*. Following the decoupling between gravity and the other interactions, gravity is treated as a property of space-time and the other interactions a property of matter/energy. For the former, General Relativity holds, for the latter the quantum paradigm is applied. At the still incredibly high densities just after the Planck era, matter/energy is described by means of a single undifferentiated field, the inflaton. Using the language of quantum fields, the lowest energy state of the inflaton is a "vacuum", but the corresponding enormous energy density drives the inflation. While expanding, at the crazy rate mentioned above, the energy density decreases. While it decreases, the different components (strong interaction, weak interaction, and electromagnetism) decouple from each other. At the end of the inflation, a sort of phase transition occurs and the original "vacuum" decays into the plurality of particles carrying the mentioned interactions.

During the inflation, the energy density is extremely high, but it can hardly be expressed in terms of temperature, since there are no particles around. In any case, the temperature decreases with expansion. However, just after the phase transition, the appearance of an immense number of different particles brings in a corresponding number of degrees of freedom, and in practice produces an increase of the temperature of the cosmic medium (which is now a sort of "superplasma"). This phase is called *re-heating*.

Let us go on. After the phase transition, the expansion continues at a much slower rate and with a different law, until the present time. The quarks, with their gluonic plasma, start forming hadrons, such as protons and neutrons. Protons on their own are the nuclei of hydrogen (H), but, as soon as the temperature allows, they combine with other hadrons to form more complex and stable nuclei, such as deuterons (one proton and one neutron) and alpha particles (He^4 nuclei with two protons and two neutrons). This primordial nucleosynthesis provides an opportunity for checking the theoretical description. Putting together: (i) the expansion rate given by the General Relativity theory, (ii) the masses of the protons and neutrons, (iii) the fact that protons are stable particles,[18] but neutrons are not, unless they are bound in a nucleus, and then concentrating on hydrogen (H) and helium (He), it is possible to evaluate the relative abundances of H, deuterium D and He^4 nuclei expected in the primordial universe. The calculated relative abundance of He^4, expressed as a percentage of mass, comes to 25 %, which agrees with observation.

We have oversimplified the story of the primordial nucleosynthesis, which provides much more information than mentioned above, but requires a not yet completely settled discussion (for instance on the abundance of D and traces of a handful of slightly heavier nuclei, like lithium Li^7 and beryllium Be^9).

As discussed earlier, after 380,000 years the temperature becomes low enough to allow the combination of free electrons with nuclei to form neutral atoms. From then on, the universe becomes transparent, and on a macroscopic scale gravity dominates. The tiny inhomogeneities that survived inflation start growing more and more, due to gravitational attraction. Aggregates of atoms appear, in which the growing density drives a corresponding increase in pressure and temperature. Temperature rises locally to a level triggering the nuclear fusion of hydrogen into helium. The *first generation* stars are born. They are comprised essentially of hydrogen with very little helium, and they start "burning" hydrogen.

Looking at Fig. 11.4, we see that the scale of the inhomogeneities is much greater than that of stars, and indeed stars, just as did atoms, begin to cluster into vast conglomerates, which are the primordial galaxies.

At this point, the evolution times are no longer expressed in fractions of a second, but rather in millions or even billions of years. Stars, growing older, convert their hydrogen into deuterium and helium and, depending on the initial mass, may ignite the fusion of helium in the stellar core. If the mass is large enough, the temperature and pressure in the core may reach values sufficient to trigger the fusion of heavier nuclei, up to iron. The energy released during the nuclear fusion process is responsible for a strong radiation pressure inside the star, opposing and counterbalancing the weight of the outer layers of matter. However, for very massive stars, the increase of the central pressure and temperature continues until the start of the synthesis of nuclei heavier than iron.

These nuclear processes are now energy demanding, so that the radiation pressure begins to decline, no longer supporting the weight of the outer layers of the star. In a very short time the core collapses. The resultant rapid energy release cannot stop the collapse, but blows off the outermost layers of the star, giving rise for days, or even weeks or longer, to an abnormal brightness. A *supernova* has

---

[18]Some theories consider the possibility of a decay of protons too, but it would happen in times longer than the age of the universe.

appeared. The outer layers of the star, containing traces of heavy nuclei corresponding to the elements we find even in the crust of our planet, are blown away and dispersed in space, while a central remnant is left either in the form of a *neutron star*[19] or of a *black hole*.[20]

We cannot enter here into a description of the amazing variety of different objects, from the scale of galaxies down to planets and just plain "celestial rocks". This is the domain of astronomy and astrophysics. What is relevant is that the whole scenario we have outlined, in its essential lines, gives a consistent account of the history of the universe, although grey areas and open puzzles remain which require further investigation.

11.8 Old and New Problems

The scenario described in the previous section is a powerful conceptual framework, but leaves behind some open problems and unexplained features. First comes the already mentioned fact that we are unable to describe the physics of the Planck era, where the two main *cathedrals* of contemporary physics, general relativity and quantum mechanics, exhibit an unsettled mutual inconsistency. Physics is always in trouble when singularities appear, as in the case of *black holes*. Black holes are predicted by the theory, but scientists cannot understand the condition of matter within them, or in the immediate surroundings of the central singularity.[21] Yet the general belief is that black holes do indeed exist. We have, for example, evidence of a giant black hole at the centre of our galaxy (and suspect the presence of giant black holes in the centre of most galaxies). In our galaxy the black hole is called Sagittarius A*, and has a mass of more than 4 million solar masses.

Besides singularities, however, there are other puzzles, which we discuss in the following.

11.9 The Missing Mass

Missing mass was the expression used by Fritz Zwicky in the 1930s, when analysing the orbital speeds of galaxies in clusters. Apparently the velocities were higher than expected for objects tied to each other by gravitational forces. These

---

[19]A typical object having a radius in the order of a few km and the density of an atomic nucleus.

[20]A fascinating and partly mysterious object predicted by General Relativity, where gravity is so strong that nothing, not even light, can escape.

[21]The problem is "solved" thanks to the presence of the "horizon": the last surface from which something can escape. Below the horizon, a cosmic censorship principle is acting: we cannot see in there, so why should we bother?

excess velocities were also found in the rotation of galaxies, including ours. The observed velocities could be accounted for by assuming that the galaxy, or the galaxy cluster, contained more mass than observed. Of course, in the skies there is matter in objects other than in the ones shining their light towards us, e.g. in the dead remnants of small stars, planets, asteroids, meteorites and dust. The quantity of all that obscure matter can be roughly estimated and is expected to be less than the mass of visible stars. However, in order to explain the anomalous orbital velocities, the missing mass in the universe must be more than five times the visible mass.

Today scientists no longer use the term "missing mass", but rather refer to the *dark matter* content of the universe. The remarkable thing is that, just by assuming the actual existence of dark matter, everything can be explained within the framework of general relativity. Unfortunately however, we do not know what that dark matter could be. There are many theories conjecturing types of matter that do not interact with the electromagnetic field. Neutrinos and/or other more or less exotic particles have been proposed, but the corresponding theories lack direct experimental evidence, are inconsistent or simply do not work. Another possibility, of course, is that general relativity is somehow incomplete or even wrong. Here too, many theories have been put forth, but often they raise more problems than the ones they solve.

11.10   The Foamy Distribution of Galaxies

One hundred years ago the universe was thought to be an infinite homogeneous stellar field. Then it was recognized that the stars are grouped into galaxies, so the universe became an infinite homogeneous distribution of galaxies. Since the mechanism leading to the formation of stars, then of galaxies, is the action of gravity starting from accidental density fluctuations, it is easily acceptable that galaxies also group in clusters, and these clusters in superclusters. In any case, enlarging the scale, we expect the universe still to be homogeneous.

However if we plot the global distribution of galaxies starting from the Earth as an origin, and use Hubble's law in order to determine the distances from us, we finally obtain Fig. 11.5, in which each little dot is a galaxy and the colours are used to mark distances (blue means closer and red is farther). The impression one gets is that of a spongy or foamy structure.

Apparently galaxies are located on enormous walls and filaments bordering immense bubbles, or *voids*, where nothing is visible. Why is it so? It may depend on what happened before the recombination era and/or on the distribution of dark matter, but we really do not know.

Fig. 11.5 Distribution of galaxies across the sky, as seen from the earth. The picture is taken from the Two Micron All-Sky Survey (The source of the picture is http://www.ipac.caltech.edu/2mass/gallery/showcase/allsky_gal_col/index.html. The authors are 2MASS/T.H. Jarrett, J. Carpenter, and R. Hurt. The file is available on Wikimedia Commons.)

11.11   The Accelerated Expansion

In the title of a previous section we mentioned a date: 1998. We can explain now the reason for this choice. At the end of 1998 and at the beginning of 1999, two independent groups of astrophysicists, guided respectively by Adam Riess and Saul Perlmutter, published the results of their extended surveys of *type Ia supernovae* (*Sne Ia*) located in other galaxies. Remember that supernovae are massive stars which at a given stage of their life implode, and in the process release enormous amounts of energy in a short time. There are various types of supernovae; among these is type *Sne Ia*. Due to the peculiar mechanism that leads to their implosion, and the initial masses of the original stars, these supernovae produce a *light curve*[22] which is almost the same, independent of their location in space and time. This feature makes them convenient *standard candles* for use in distance measurement. Since their intrinsic luminosity is given and fixed, their apparent luminosity tells us how far away they are.[23]

Supernova explosions are rare events; about three occur per century per galaxy. They are however extremely luminous events, so that it is possible to observe them in other galaxies. As we have seen, a supernova's light curve gives us the distance to the supernova and thus to the galaxy containing it. However, we can also infer this distance from the observed red-shift using Hubble's law. The red-shift can be obtained from one of the spectral lines visible in the light emitted by the supernova.

---

[22]The light curve is the diagram of the light intensity released in time.

[23]We have discussed this form of astronomical distance measurement in Chap. 3.

The unexpected result is that for type *Sne-Ia* supernovae at distances larger than approximately 1 billion light years, discrepancies between the distances obtained from the red-shifts and the light curves emerge. The supernovae appear to be fainter than they should. Since both types of data come from the same sources, the explanation cannot be in terms of different distances.

One may conclude that the expansion rate of the universe at the time of emission of the light (measured by the red shift) was less than it is today, so that the path length travelled by the light (on which the apparent luminosity depends) was longer than one would have expected from the initial expansion rate. More explicitly: the expansion of the universe appears to be accelerating.[24]

In fact the FLRW equations admit three different solutions depending on the actual value of the average energy/matter density in the universe. In all cases, the gravitational attraction between galaxies slows down the initial expansion.

In the first solution, if the density is large enough, the expansion reaches a maximum and the universe begins to contract back, ending in a final *Big Crunch*, which is the antipodal singularity of the Big Bang. This would be a *closed* universe. In the second solution, for a critical value of the density (approximately five hydrogen masses per cubic meter), the expansion slows down and asymptotically stops at an infinite cosmic time. This would be a flat and *open* universe. In the third solution, for smaller densities, the expansion rate slows down too, but tends to an asymptotical value different from zero, so that the expansion never ceases. This would be an *open, but not flat* universe.

Under no circumstances do the FLRW equations yield a solution where there is an acceleration of the expansion rate. So what is the driver of the observed accelerated expansion?

11.12 The Flatness of Space

The difference between the three cosmological solutions of the FLRW model may be expressed in terms of the curvature of space-time and in particular of the curvature of space. In practice, a curved space means that its geometry is non-Euclidean, i.e. the sum of the internal angles of a triangle is not 180°. Astronomical observations over cosmic distances and an analysis of the inhomogeneities of the CMB apparently tell us that space in the universe is, on the average, *flat*, i.e. Euclidian. This corresponds to the intermediate FLRW solution, which requires a *very precise* and critical matter/energy density. This *fine tuning* is disturbing because it is surprising that any physical quantity would assume a critical

---

[24]The Nobel Prize in Physics in 2011 was awarded to three members of the two teams making this discovery.

value, such that the smallest change will induce the universe either to collapse eventually, or to expand for ever at a finite expansion rate.[25]

However, the main problem is that visible matter makes up less than 5 % of this required critical density. If space is indeed flat and if we trust GR, there has to be *something else* around and it has to be as much as 95 % of the total. We have already cited the problem of dark matter, required to explain the dynamical behaviour of star clusters, galaxies and galaxy clusters. According to the estimates deducible from the data of the WMAP[26] observation mission of NASA, the ordinary matter density is 4.6 % of the critical density and the dark matter density is another 24 %. Together they make up 28.6 % of the required amount, so 71.4 % is still missing. Furthermore, unlike both ordinary matter and dark matter, this 71.4 % must be something that does not gravitate, but which aids the flattening of space. Since it produces physical effects, without being "matter", this mysterious ingredient has been named *dark energy*.

11.13   The Concordance Model of the Universe

At the end of last century, the accepted view of the universe converged towards what is now called the *cosmic concordance model* (also known as the cosmic standard model), where consistent explanations were given for most of the puzzles arising from the observations.

The concordance model is also designated by the acronym ΛCDM (Lambda Cold Dark Matter). The first relevant assumption of the ΛCDM model is that the framework within which all observed phenomena should be explained is General Relativity. As mentioned above, GR does indeed work, provided that one assumes the presence in the cosmos of additional forms of energy and matter. Both in GR and in quantum field theory, one most often follows the Lagrangian approach mentioned in Chap. 4. To preserve the GR framework, it is sufficient to introduce into the *Lagrangian* of the universe appropriate additional terms, beyond the usual ones for baryonic matter.

Let us then start with the "Λ" appearing in the name of the model. Λ is indeed nothing other than the good old cosmological constant that Einstein dubbed as the biggest blunder of his life. Cosmologists have verified that simply introducing that constant into the equations accounts quite well for the accelerated expansion of the universe. The mathematics is thus satisfied; however physics is left with the burden of finding a suitable interpretation for Λ. We have already mentioned its interpretation as dark energy accounting for 71.4 % of the content of the cosmos, and that it

---

[25]There exist a number of physical quantities, in which small changes would preclude the existence of the world (and life) as we know it. We will discuss these further in Chap. 13 in the section: The Anthropic Principle.

[26]Wilkinson Microwave Anisotropy Probe.

does not produce gravity (as it is not a form of matter), but rather generates a negative pressure responsible for the accelerated expansion of space. We may think of it as being a fluid permeating the universe, but its density always stays the same (i.e. it is a *real* constant!). Normally we would expect that in an expanding space any fluid progressively dilutes, but this is not the case for Λ. In short, we do not know what the dark energy is, but it works!

The remainder of the initials in ΛCDM come from Cold Dark Matter. We have already seen that some dark matter is needed to account for the observed effects of gravity in large scale clusters of matter. Theorists have put forth various hypotheses about the nature of DM and have proposed many candidates. The different proposals may be grouped into three categories: *cold, warm and hot dark matter*. The three categories have to do with the free-streaming length[27] of the particles, or to their speed: *cold* means slow, i.e. non-relativistic; *warm* means weakly relativistic and *hot* means relativistic. All three types are expected to interact only very weakly with ordinary matter, but each have different influences on the size of the structures (galaxies, galaxy clusters, and superclusters) appearing in the universe. The observed distribution and size of structures is reasonably consistent with the dominance of cold dark matter. Hence it has been included in the concordance model.

We may summarise the description of the evolution of the universe given by the ΛCDM model by means of the graph represented in Fig. 11.6,[28] where space-time appears as a bell-shaped hyper-surface.

The image is inspiring but must be handled with care. The surface is necessarily represented as bi-dimensional (being drawn on a paper sheet), but actually is *three-dimensional*. The third dimension of the figure is time. Everything starts on the left with the singularity; close to it, what happens is dominated by the interaction of strong gravity with quantum fluctuations. Then follows an inflationary exponential expansion; in a tiny fraction of a second the scale of the universe grows from microscopic to cosmic.

As the temperature of the content of the universe decreases, the cosmic nucleosynthesis begins. The nuclear plasma cools down and the expansion rate of space slows down. At the cosmic age of 380,000 years, (neutral) atoms are formed and the cosmic microwave background originates. A "dark age" follows, during which atoms start to coalesce around positive fluctuations of the matter density, thus increasing the local temperature and pressure. Finally the temperature and pressure into the cores of the agglomerates become large enough to ignite nuclear fusion, and stars begin to be born. Once more light is emitted in the universe. Stars group into galaxies, clusters, superclusters, and filaments, etc.

---

[27]The free-streaming length is the average distance a particle can travel without interacting with matter scattered along the way. The weaker the coupling with ordinary matter, the longer the free-streaming length.

[28]This image was created by the NASA WMAP Mission and is downloadable from Wikimedia Commons at https://commons.wikimedia.org/wiki/File:CMB_Timeline300_no_WMAP.jpg.

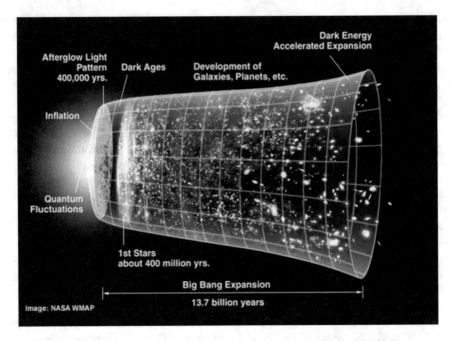

Fig. 11.6 Pictorial view of the evolution of the universe. The bell-shaped surface (which should be three-dimensional) is the border of the visible universe; cosmic time is measured along the surface starting from the initial singularity

Until now, the matter/energy density has been large enough to allow gravity to act as a brake on the expansion, whose rate has been steadily decreasing. However, as the expansion progressively dilutes the matter density, dark energy (that does not dilute) takes over and converts the expansion from a decelerated to an accelerated stage, which is the phase in which we are now living. Transverse sections of the bell shown in Fig. 11.6 are circles, but in the real world they are spheres, each of which corresponds to the frontier of the visible universe.

11.14 Alternative Scenarios

The strength of the concordance model is in its relative simplicity and internal consistency, but it does not account for everything, and there still remains that 95 % of the universe designated as "dark", whose physical nature is unknown. This situation has stimulated a continuous effort to look for alternative solutions.

Let us start from the dark energy in the form of the cosmological constant. A vast literature exists presenting attempts to find more intuitive or better founded versions of the dark energy. Often it has been described as a field or fluid, whose density is not really constant, despite the expansion of the universe. The names used to designate the candidates for these versions of dark energy are sometimes whimsical: e.g. *ghost energy* or *quintessence* (from Aristotle's "fifth essence", the *ether* that filled up the skies). Leaving names aside, most theories commence by manipulating the Lagrangian of the universe and introducing new terms tailored in an ad hoc fashion to reproduce the desired properties.[29]

Other approaches attempt to find analogies between dark energy and known phenomena. An example worth mentioning is the *Strained State Cosmology* [1], where the dark energy is treated as the deformation energy density of a four-dimensional elastic continuum. In fact we know that in three dimensions, when trying to deform an elastic material, a strain energy density is stored in the continuum. That strain energy is related to the reaction and resistance of the material against the deformation. In practice the accumulation of strain energy hinders the deformation and tends to recover the unstrained state. Extending the analogy to four dimensions, if we accept the idea that space-time is not just a mathematical artefact, but has physical properties of its own, a strain energy density is associated with the presence of curvature and it works against it, driving an accelerated expansion towards an asymptotic unstrained and completely flat state. This theory succeeds in explaining the observed accelerated expansion and also the exponential expansion close to the origin.

A separate class of theories encompasses all attempts to extend GR, modify it or simply set it aside. However, for the time being, the ΛCDM model remains a very simple conceptual framework accounting for almost all the observations, though at the price of a difficulty in the interpretation of the nature and properties of its ingredients.

So we have arrived at the end of our journey through space and time, or more correctly *space-time*. What we hope to have presented in this chapter is a modern view of a puzzle that has occupied the minds of *homo sapiens* ever since the first of our kind looked up at the sky and beheld the vast expanse of stars stretching across the heavens. Not all questions have been answered, and some of today's answers will surely be rejected in the future. That is the evolutionary nature of science.

We have raised a number of issues in this chapter that warrant further discussion and speculation. We will defer this discussion to Chap. 13, our concluding chapter. In the next chapter we address the topic of Complexity.

Reference

1. A. Tartaglia (2016), *The strained state cosmology*. Int. J. Mod. Phys. A, 31, p. 1641015–1641024.

---

[29]See Chap. 3 for a discussion of the pitfalls in models.

Chapter 12
Complexity and Universality

> *It used to be thought that the events that changed the world were things like big bombs, maniac politicians, huge earthquakes, or vast population movements, but it has now been realized that this is a very old-fashioned view held by people totally out of touch with modern thought. The things that really change the world, according to Chaos theory, are the tiny things. A butterfly flaps its wings in the Amazonian jungle, and subsequently a storm ravages half of Europe.*
> Neil Gaiman, Good Omens: The Nice and Accurate Prophecies of Agnes Nutter, Witch

Abstract In the last chapters we presented some of the most glamorous advances in Physics itself. Now we are ready to discuss some of the most intriguing among the countless contributions of the physical methodologies to other sciences. New approaches have been developed, for which the emergence of new concepts, patterns, universalities and the cross-fertilization among different phenomenologies, matter more than the field of application. Some of them, such as Complexity, have been very fruitful and brought along with them a new perspective on science and the underlying reality.

12.1 Simplicity and Complexity

In the previous chapters, we have explored the Universe from the tiniest particles to the largest celestial bodies. To summarize in a few words what we have found, the dream of a Theory Of Everything (TOE), which has been pursued for more than 2000 years by philosophers and scientists alike, seems to be as remote as ever, in spite of the enormous progress that physics has achieved in the last two centuries.

Just as if we were climbing a mountain, we find that the more we go up in our learning, the more the horizon of our ignorance opens up. And yet, in spite of the scope and complexity of what remains unexplored, simple laws and patterns have emerged to illuminate us and let us be reasonably sure that we are on the right track, in spite of the occasional detour. The antithesis between the simplicity of the basic laws of nature (e.g. the unification of the fundamental forces or the elegance of relativity) and the awe-inspiring complexity of the world around us is indeed

© Springer International Publishing Switzerland 2016   179
R. Barrett et al., *Physics: The Ultimate Adventure*, Undergraduate Lecture Notes in Physics, DOI 10.1007/978-3-319-31691-8_12

puzzling: complexity and simplicity seem to be like two lovers, who are profoundly different, yet complementary and in constant need of each other.

The roots of complexity are self-evident: in Chap. 7 we have remarked that in Newtonian gravity it is possible to find exact analytical solutions only to the two-body problem, and in General Relativity not even that is feasible. Of course numerical solutions abound even for a large number of bodies and, with the present computational power, they can be extremely accurate. However, according to recent estimates, there are 10^{24} stars and the number of particles present in the Universe exceeds 10 to the power 80! Can anyone even catch a glimpse of what such numbers really mean?

Descending to more earthly matters, how many snow particles can result in an avalanche and how can we predict when it will happen? Or, how many elements of perturbation can affect weather predictions on a local scale, or the sudden onset of an earthquake, or of a volcanic eruption? In the presence of such numbers, the search for reliable solutions (and consequently predictions) seems to be totally hopeless, particularly since—although Neil Gaiman's statement above is almost absurdly exaggerated—Chaos theory teaches us that a very tiny event may indeed ultimately be the cause of a huge unexpected effect.

Certainly statistics may be very useful, but if I plan a hike for tomorrow, how can it help me to know that there is a 50 % probability of rain? And then, in spite of the huge computing power, even the best weather forecasters sometimes fail us. No economist can reliably predict the stock market fluctuations, no geologist can accurately predict the occurrence and magnitude of earthquakes, and even the behaviour of close friends and relatives can often be totally baffling. All this is complexity. Very confusing, yet a world without complexity would be very boring indeed.

Making sense out of the world complexity is, and has always been, the (un-stated) purpose of every philosopher and scientist, and we have reviewed through the various chapters the contributions of physicists to this goal. However, in the last few decades we have seen a remarkable new approach to the problem, with a windfall of very important results in a broad interdisciplinary context. This will be the focus of the present chapter, but first we wish to discuss the meaning, from a physicist's point of view, of the two words *simplicity* and *complexity*.

Let us start with a short story to illustrate what *simplicity* is not:

A father and son are walking together in the country. Suddenly the child enquires why the sky is blue. Very seriously the father explains: "It is blue because God thought that blue is a beautiful colour". "And why is the grass green?" "Because, as you can see, blue and green fit well together". *The son goes on with other questions and the father keeps answering in a similar way. But after a while he becomes concerned and tells the boy*: "Please let me know, if I am getting too technical".

The joke lies, of course, in the antithesis between the father's concern and his simpleminded explanations. But are his explanations really simple? Not for a sci-entist, because the first answer implies the existence of a God, and since so many religions claim to be the only veritable source of truth about God, the question arises: which of them, if any, is the right one? Under the surface of an apparent simplicity are hidden some of the most difficult eschatological questions, to which a scientist as such cannot find an answer. However, even a physicists' answer[1] to the question: "why is the sky blue?" is far from being exhaustive. In fact, under a critical analysis, one finds that, although no transcendental questions arise, the answer is still unsatisfactory, since most of the arguments do in turn require deeper explanations, and there is no end to the process.

To make things worse, it is not even simple to define what *simple* (or complex) means, partly due to the many differently nuanced meanings we commonly give to words in different contexts and languages. The Nobel laureate Murray Gell-Man, who, as we saw in Chap. 10, is best known for his elegant (or *simple*) classification of the elementary particles, starts by defining, for a given entity, its *effective complexity* (or conversely *simplicity*) as the length of a highly compressed description of its regularities. Here *compressioncompression* means the elimination of redundancies, which would otherwise make the description arbitrarily long. He goes on to coin a name "plectics", derived from the fusion of the two words *simple* and *complexity,* for the new branch of science, whose goal is the study of complex systems, which is of paramount importance at the Santa Fe Institute.[2]

12.2 Complexity Theory

Avalanches, earthquakes, the weather, the stock market or the behaviour of a large crowd are just a few examples of complex systems. In biology or medicine the complexity of organs and organisms is self-evident. Although a universal definition of Complex Systems (CS) does not exist, a common denominator among them is the difficulty of predicting their behaviour reliably, due in most cases to the extremely large number of constituents (or components).

In most cases, even assuming it would be feasible to assemble all the necessary data on time, an exact and detailed formulation of the problem is not possible, much less so a solution. Hence statistical techniques are often deployed, in many cases yielding satisfactory results. However, as in the previous example of weather monitoring, if you wish to predict the occurrence of an earthquake in a given location, the knowledge that you can expect one in the next 20 years does not help

---

[1]Sunlight scatters off molecules of gas and other small particles in the atmosphere in a random process called Rayleigh scattering. It is this scattered light that gives the sky its brightness and colour. Since Rayleigh scattering is inversely proportional to the fourth power of wavelength, shorter wavelengths, such as violet and blue scatter more than longer ones, such as yellow and red. However, the whole process is much more complex, since this elementary explanation does not take into account other factors, such as the solar spectrum and the absorption properties of the Earth's atmosphere.

[2]The Santa Fe Institute is an independent, nonprofit theoretical research institute located in New Mexico and dedicated to the multidisciplinary study of the fundamental principles of complex adaptive systems. It was founded in 1984 by Gell-Man and a few other scientists, mostly from the Los Alamos National Laboratory.

very much. More useful results can often be obtained by modelling the CS^3—i.e. finding a simplified representation of it, in order to reduce it to a more tractable system, with solutions approximating those of the original problem—and via numerical simulations, in which *numerical experiments* are performed.

Since the components of CSs are interconnected and at least locally interacting, networking is another important aspect of Complexity Theory (CT). In addition most, if not all, CSs are nonlinear, hence nonlinearity is another essential ingredient to study. Linearity, i.e. expecting the effects to be proportional to the causes, is a naturally ingrained tool in our brain, since it allows us to take quick decisions. It has, however, in most cases only a limited range of validity. With the power of modern computers it has become fully possible to tackle nonlinearities reliably, and the results are indeed very exciting. A merely linear world would be quite uninteresting.

The relevance of modelling, simulations, networking, nonlinearity and other concepts and techniques of CT far transcends the boundaries of physics. For what concerns its applications, from the examples mentioned above we can easily surmise that complex systems can be found in practically all fields, from physics, geology and biology to economics, social sciences and philosophy. Actually, lest the reader be led to think that CT is applicable to *all* scientific problems, let us make a couple of counterexamples, viz the classical 3-body problem and the analytical solution of a fifth order algebraic equation. Both problems are not only very difficult, but actually insoluble. Yet neither lies within the scope of CT. The former has already been mentioned several times before; the latter involves some curious incidents from the history of mathematics.[4]

---

[3]See Chap. 3 for a discussion of modelling.

[4]Every high school student can solve second order algebraic equations. However, third order equations represented an unsolved enigma up to the early 16th century, when the mathematician Scipione del Ferro found a solution for a special case of them. At that time mathematical solutions of relevant problems were not usually published, since they allowed their owner to challenge other mathematicians and make money in highly publicised *duels* (in which problems were proposed and whoever could solve more of them would be the winner). Before he died, del Ferro confided his secret to his student, Antonio Fiore.

A young mathematician, Niccolò Tartaglia, challenged Fiore to a duel and unexpectedly won, gaining wide fame. But another famous mathematician, Girolamo Cardano, succeeded in inducing Tartaglia to reveal the formula to him by enticing him with many promises, one of which was to keep the secret. However, a few years later, Cardano published it, so that it has since been known as *Cardano's formula*. The unfortunate Tartaglia challenged Cardano to a duel, which, however, took place against a student of Cardano, Lodovico Ferrari, who in the meantime had also learned to solve fourth order equations, and consequently won.

Centuries later, Niels Henrik Abel and Paolo Ruffini proved, in their *impossibility theorem* (1823), that there cannot be an analytical solution for fifth order equations. There is, however, nothing magical in this. In fact *solution* here means simply expression in terms of standard elementary functions, such as square roots, logarithms and trigonometric functions. However, in order to obtain a square root, one needs a numerical procedure with iterations, and an exact value is not possible, since, being an irrational number, it would require an infinite number of digits. Likewise, fifth order algebraic equations are also numerically soluble to any desired accuracy by means of iterations. Introducing new functions, besides the standard ones, it would be equally possible to express these solutions analytically.

Since CT is a relatively new topic, there are many different ways to approach it, depending on our own individual taste and interest. Most natural, of course, is a computational CT approach, in accordance with a famous quote by James Clerk Maxwell: "All the mathematical sciences are founded on relations between physical laws and the laws of numbers, so that the aim of exact science is to reduce the problems of nature to the determination of quantities by operations with numbers."

As a specific instance of CT, let us consider one of the examples mentioned earlier, i.e. let us imagine that a large amount of snow has accumulated on the side of a mountain overlooking a village. Everybody knows that sooner or later the snow will fall down into the valley, but nobody can reliably predict when and/or, most importantly, whether it will be a single catastrophic event or whether it will break down into several smaller avalanches of minor consequence. It may be rather frustrating to say that CT, in spite of all its achievements, has *not* really succeeded in solving this absolute type of problem. It has, however, been very successful in solving problems of a relative type, such as establishing relations between different kinds of complex phenomena. This, of course, is of little comfort to the villagers threatened by the avalanche, yet a *cross-fertilisation* may often be extremely useful, e.g. in helping us to find approximate solutions, which might have already been obtained for related CSs.

In fact a given problem, intended as the complete information about a CS (in our case all details pertaining to the snow accumulated on the mountain and the meteorological conditions) and its solution (i.e. all the details of the consequent event, the avalanche) are just two different representations of the same physical reality, just as an equation (e.g. $3x = 6$) and its solution ($x = 2$) may be considered as two equivalent mathematical representations of the same mathematical reality. Hence the usefulness and relevance of any procedure that helps us to establish relations between different representations is immediately apparent.

Leaving aside for the moment the difficulty of predicting the outcome of extremely volatile events such as avalanches, let us now survey some of the age-old concepts upon which CT has cast a new light, providing some intriguing new viewpoints, e.g. criticality, randomness, knowledge and learning, game theory, and data interpretation. It may be interesting to remark that the latter has been replaced in many applications by a new discipline, *data mining,* due to the over-abundance of available data, which yield to the researcher a potentially rich field to harvest. One should, however, be careful to remain unbiased in exploiting these data and not forget the old adage: "If you torture your data long enough, they'll confess", with the double-entendre that they might "confess" what you want, rather than the truth.

Later on in this chapter we will also discuss some entirely new concepts, such as fractality, which have emerged out of CT. In fact a very important result of CT has been to show that similar relatively simple patterns and phenomenologies often emerge from entirely different complex systems.

12.3 On the Edge of Chaos

An example of an emerging pattern is provided by complex adaptive systems, or *self-organized criticalities* (SOC), in which order emerges at the edge of chaos, i.e. out of a system whose components are unable to lock themselves into a stable organised state, but yet retain some recurrent patterns containing recognizable persistent structures. A simple (classical) example can probably illustrate this concept better than a lengthy general description or definition:

Let us assume that somewhere there is a completely isolated island, in which only two animal species can be found: lions and gazelles. For their coexistence it is necessary that a certain equilibrium be reached, in which neither species runs too fast, for if lions prevail, they will, after a given time, eat all the gazelles, and then they in turn will all die of hunger. Conversely, if the gazelles are too fast, the lions will be unable to catch them and will eventually starve, but then the gazelles will also become extinct, because they will multiply to the point where the grass and leaves will no longer suffice to nourish them. The number of lions and gazelles will not, of course, remain constant at the equilibrium level, but will keep oscillating within a certain range around some critical values.

Such an island obviously does not exist—any ecosystem is clearly far more complex—but the conclusions can be generalised to systems with any number of different interacting players. Actually, the more players, the easier it is for the system to fall into a state of *organised criticality*. And what happens if for some reason one of the players becomes predominant, i.e. too efficient or fast, as in the island of lions and gazelles? According to the relevance and speed of the change, either a transition to a new SOC or a catastrophe occurs. An example of the former has been the reorganization of society after the French Revolution. A tragic example of the latter could be the fate of mankind if the current race towards more and more growth continues unabated and uncontrolled [1].

CT scientists endeavour, often by means of modelling and simulations, to understand how order can emerge from chaos in complex non-linear systems, both in the physical world (micro and macro) and in social contexts, such as in crowd behaviour or the stock markets. Waldrop [2] provides an almost poetical description of such a transition:

> The balance point – often called the edge of chaos – is where the components of a system never quite lock into place, and yet never quite dissolve into turbulence either ... The edge of chaos is where life has enough stability to sustain itself and enough creativity to deserve the name of life. The edge of chaos is where new ideas and innovative genotypes are forever nibbling away at the edges of the status quo, and where even the most entrenched old guard will eventually be overthrown. The edge of chaos is where centuries of slavery and segregation suddenly give way to the civil rights movement of the 1950s and 1960s; where seventy years of Soviet communism suddenly give way to political turmoil and ferment; where eons of evolutionary stability suddenly give way to wholesale species transformation. The edge is the constantly shifting battle zone between stagnation and anarchy, the one place where a complex system can be spontaneous, adaptive and alive.

12.4   Fractality

Another basic concept of CT is fractality, i.e. the tendency, shared by countless natural and man-made phenomena, to reproduce their structure at any scale (self-similarity). If the reader is not acquainted with fractality, we invite him/her to take a piece of paper, draw a line and divide it into three equal parts. The next step is to remove the middle segment and replace it with the other two sides of an equilateral triangle. This procedure can be repeated indefinitely on all the remaining segments, creating graphs, whose geometry is totally new. The graph obtained as described, but starting from an equilateral triangle instead of a straight line, is known as the "Koch snowflake"; Fig. 12.1 illustrates the first four iterations of the procedure to generate it.[5]

Another very interesting example of a fractal graph is Peano's plane-filling curve, several iterations of which are shown in Fig. 12.2.[6]

Here we have a curve (i.e. a one-dimensional entity), which asymptotically (i.e. when the number of iterations grows indefinitely), tends to cover an entire

Fig. 12.1 Generation of a Koch snowflake

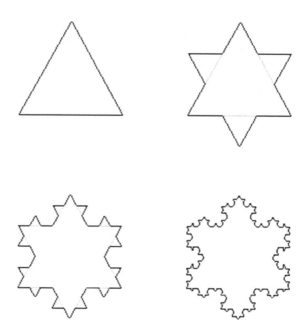

---

[5]This image has been uploaded to Wikimedia Commons and made available under the Creative Com-mons Attribution-Share Alike 3.0 Unported license (https://commons.wikimedia.org/wiki/File:KochFlake.png).

[6]This image has been uploaded by António Miguel de Campos to Wikimedia Commons and made available under the Creative Com-mons Attribution-Share Alike 3.0 Unported license (https://commons.wikimedia.org/wiki/File:Peanocurve.svg).

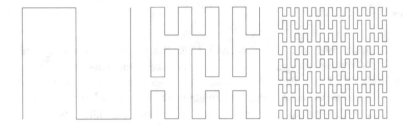

Fig. 12.2 Generation of a Peano *curve*

two-dimensional surface. How is it possible that its dimension changes from 1-D to 2-D, and what does this mean?

Many other examples of graphs exhibiting fractality could be presented here, together with the corresponding analytical discussions. Yet, for many decades, the concept of fractality remained essentially confined within the realm of mathematical speculations about problems such as differentiability, proofs of existence and uniqueness, topology, number theory and others, of interest only to mathematicians (except perhaps as curiosities). All this changed in the sixties, when Benoît Mandelbrot coined the word "fractal" and developed fractality as a new interdisciplinary field of research and applications. The striking computer-generated visualisations which accompanied his research papers and books helped to popularise his work and turn him into the widely acclaimed father of *fractal geometry*.

In the new field—which could also be nicknamed *the geometry of roughness*—the idealised Euclidean world of perfect shapes, such as points, lines and circles, is replaced by hopelessly irregular shapes, such as coastlines, clouds, blood vessels, tree shapes, or even plots displaying the daily fluctuations of the stock market. Amid all this roughness and variety, some peculiar properties emerge, such as the already mentioned self-similarity, and Mandelbrot's lavish illustrations had the power of conveying the message.

However, if the popularity of his figures helped him to defeat the criticism of his opponents, it was the richness of applications in all fields, from exact sciences to econometrics, plus the cross-fertilisation arising from the emerging similarity between totally unrelated phenomenologies, that made fractality (in particular) and complexity (in general) so fruitful. Fractality is also an instance of the application of a very general *conjecture*, which we wish to discuss in the next section.

12.5   An Epistemological Conjecture

As we have already mentioned in Chap. 1, mathematics is *not* a natural science, since it is totally a priori, i.e. independent from our experience of the world external to us (including our body, but not our mind). In fact, one could be blind and deaf and still be able to develop inside one's mind some mathematical theories, although one would hardly be motivated to do so. On the other hand, science, indeed all natural sciences, are a posteriori, in the sense that they must start from observational or experimental data. Nevertheless, as we have stressed several times in our book, mathematics is a basic tool for the development of science.

In our opinion, there is an additional link between mathematics and science, which we cannot really prove and therefore propose as a (cosmological) *conjecture*[7] (whose general applicability is, however, very questionable):

> Every major mathematical breakthrough is accompanied, sooner or later, by the discovery of some phenomenology based on it.

In other words, there seems to be a bond between mathematics and natural science, or more specifically, between the a posteriori and the a priori realms. Hence the two approaches advance in parallel, although at different paces: mathematics is not only the language of science, but it seems also to make up the bones and structure of it.

Our formulation of the conjecture is, of course, scientifically unacceptable, since it uses words, such as *major*[8] and *sooner or later,* which are not properly defined in the context. Also, a bit of humility should warn us that, as a species born and evolved on a little planet orbiting an average star among zillions, we should not expect to be able to make such general statements. Nevertheless we may certainly use the conjecture as a working hypothesis to guide us in this tremendously complex world, which we endeavour to explore.

In fact, as cited in the Preface of a book devoted to universalities [3], *variety is perhaps the most amazing attribute of Nature, with an almost endless array of different molecules and tens of millions of different forms of life. Yet, in spite of this bewildering diversity, there are some common patterns that are found over and over again in completely unrelated contexts.* These patterns are called *phenomenological universalities (PUN)*, and they owe their common origin to some mathematical *niche*, consistent with the above conjecture. They will be discussed in detail in the next section.

12.6 Phenomenological Universalities (PUNs)

In order to give a formal definition of PUNs, let us recall an old definition of "integers" in mathematics [4]: integers may be defined as the 'Inbegriff' (i.e. the inclusive concept, the quintessence) of a group of objects, when their nature is

---

[7]A conjecture is defined in the Merriam-Webster dictionary as an opinion or idea formed without proof or sufficient evidence. More specifically, in mathematics, as a proposition before it has been proved or disproved.

[8]Actually we would have preferred to use, instead of "major", a word conveying our profound admiration, such as *beautiful,* but we have refrained, in order not to get involved in another *trap,* i.e. the necessity of its definition in our context.

completely disregarded. Likewise, PUN's may be defined as the 'Inbegriff' of a given body of phenomenology, when the field of application and the nature of the variables involved are completely disregarded. Since we must disregard the field of application, the boundaries of traditional disciplines must also become blurred. In other words, just as we are moving, whether we like it or not, towards a globalisation of world affairs, the same thing is happening in science.

It may, however, be useful to illustrate the concept of PUNs with a concrete example. Let us then focus our attention onto growth phenomena, which represent a very important and general class of problems in all fields of science: e.g. the growth of crystals in physics, of animals in biology, of tumours in medicine, of prices in economics, etc. Note that the "growth" may also be negative, or at a variable rate (in which case it is more properly called evolution). It would appear that each case should be studied individually, i.e. by hand tailoring a model within the framework of the discipline to which it belongs, thereby transforming it into an equation or a set of equations, and searching for a suitable method to solve it.

However, the considerations of the previous section suggest a more efficient procedure, i.e. searching for a mathematical niche, to which all the above mentioned phenomenologies [5] (and many others) are related. Such a niche can be found in a Taylor expansion (see Chap. 2) associated with the evolution function under very general assumptions, independently of the field of application [6]. Once more, for brevity and simplicity, instead of a formal presentation we limit ourselves to the consideration of a particular example, i.e. the growth of a tumour. In this case the evolution function is the mass m(t) of the tumour. As in most similar problems, we can begin by assuming that the growth in unit time dm/dt is proportional to the current mass. This basic relationship yields a differential equation

$$\frac{dm}{dt} = \alpha m$$

whose solution is an exponential function: $m = e^{\alpha t}$, as can be readily verified by differentiating both sides of the solution with respect to time. The number "e", called Euler's constant, is very important in mathematics, where it is second in importance only to π, but here its choice is irrelevant.

This solution, $e^{\alpha t}$, tells us that at the very beginning (stage 1), the tumour grows exponentially. Such a rate of growth is faster than a growth at *any* power of t (e.g. $t^2, t^3, \ldots$ or t^{100}). However, this rate is unsustainable: if left unchecked, the tumour would soon become larger than the entire body. There must therefore be some biological mechanism to slow it down, and that mechanism is the development of a necrotic (dead) core inside the tumour. All the nutrients (in the blood) are absorbed from the outer shell of the tumour itself, so that the poorly irrigated core begins to starve (stage 2).

At this point, however, in order to continue its growth, the tumour begins its counterattack (stage 3) by means of several mechanisms, one of which is *angiogenesis* (generation of capillary blood vessels to reroute some blood into the core) and another is *metastasis* (diffusion of tumour cells elsewhere in the body).

These are (unfortunately) very common clinical observations. However, what is really intriguing from our point of view here is that these three stages are nothing but the first three terms of iteration in the Taylor expansion corresponding to the growth model we have considered (see Fig. 12.3). Of course, we are interested in the growth kinetics of tumours in vivo, i.e. in living patients, but the invasive techniques which would be required to collect the relevant data are clearly unacceptable. Hence we must be satisfied with a biological model.

A convenient experimental tool that captures the most relevant features of the kinetics of tumour growth are the multicellular tumour spheroids (MTS), to which Fig. 12.3 refers. MTS are spherical aggregations of tumour cells that may be grown under strictly controlled conditions. Their simple geometries and the ability to produce them in large quantities have led to interesting new insights into cancer research.

In Fig. 12.3 the triangles refer to experimental data [7] while the squares and circles are obtained by means of mesoscopic simulations (totally unrelated to the PUNs) [8]. Of course mathematics will never give you a clinical interpretation of what is happening, but it does provide you with the quantitative outcome, i.e. the predicted growth curves. Equally amazing is the fact that these curves are exactly the same for countless other phenomenologies (growth of crystals, animals, prices, etc.), even if the actual numbers involved are completely different. The iteration could certainly continue beyond the third stage, but, surprisingly, nobody (to our knowledge) has so far done it.

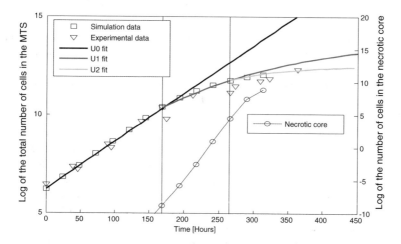

Fig. 12.3 The three phases of growth of MTS's. Temporal evolution of an MTS made of EMT6/Ro mouse mammary carcinoma cells grown in a confined culture medium. The experimental data (*triangles*) are taken from Ref. [7]. The "squares" and "circles" correspond to the total numbers of MTS cells and necrotic cells, respectively. They have been obtained from a mesoscopic simulation, based on the model of Delsanto et al. [8]. Finally U1, U2 and U3 refer to the three stages mentioned in the text

In conclusion, *universalities* can help us as an Ariadne's thread across our labyrinthine Universe, teaching us that, sometimes, the traditional bottom-up approach to science may fruitfully be replaced by a top-down procedure, as suggested by our epistemological conjecture. Also, discovering similar patterns in different contexts can help us not only to take advantage of the similarities, but also to characterize, as though with a magnifying glass, the eventual differences.

12.7   The Universality of Growth

In Chaps. 4 and 6 we saw that Newton's law of gravity and Maxwell's Electromagnetism represented two unparalleled achievements of Physics in virtue also of their *universality*. The unification of the electroweak and strong nuclear forces in the Standard Model (see Chap. 10) may be regarded as a third triumph, which however requires the inclusion of gravity in a Grand Unification Theory to be complete—assuming, of course, that this accords with the Grand Plan of Nature. Likewise, in Chemistry, Mendeleev's Table [9] owes to its *universality* (i.e. its capability to classify all the known *and unknown* elements) to both its elegance and enormous predictive power.

In keeping with the multidisciplinary character of this chapter, we may ask ourselves if it is also possible in Biology to find *universal laws*. At first sight the answer would appear to be negative, due to the tremendous complexity and diversity of life, spanning (see Fig. 1.1) almost 30 orders of magnitude in mass between a virus and a blue whale and encompassing 5 or 6 *kingdoms* (from *monera* to *animalia*). However, the use of tools borrowed from Physics, such as a mathematical reformulation of the problem and *scaling,* may lead us to a positive answer.

The concept of *scaling* has always been of basic relevance for physicists, even if often it was only considered implicitly. In his treatise of 1638 Galileo mentioned and analysed many examples of scaling [10]. On the lighter side, he is reported to have remarked that "a little dog may carry on his back two or three dogs of its size, while a horse probably cannot even carry one". Returning to the here and now, in the last paragraph we quoted an example (a virus and a blue whale) with 30 orders of magnitude difference in mass and referred to Fig. 1.1, which shows about 10 orders of magnitude difference in length between these two "animals". In writing that, we were implicitly applying scaling, since a consequence of scaling (when it is valid) is that a factor f in length corresponds to a factor f^2 in area and f^3 in volume. The same consideration applies to the concept of *body mass index* (BMI), which we mentioned in Chap. 2.

As we hinted above, there are some instances where scaling applies (or it should, e.g. for the *BMI*), in which case we call it *isometric scaling,* and other instances where it does not (such as in Galileo's *horses vs. dogs*), and then we speak of

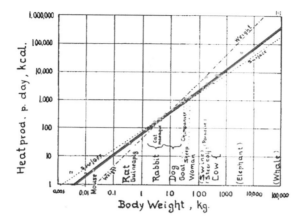

Fig. 12.4 Log of the metabolic rate versus log of body weight from Kleiber's hand-drawn figure in his classical 1947 paper. The highlighted band shows the line obtained assuming that the metabolic rate is proportional to the 3/4 power of body weight. For comparison the figure shows also the line by which the results would have to be represented if the metabolic rate were proportional to body weight or surface (2/3 power of body weight)

allometric scaling.[9] Figures 12.1 and 12.2 are totally scale-invariant, as can be immediately realized by the fact that both x- and y-scales are dimensionless. In fact, the use of dimensionless units is a typical Physicist's trick to confer more generality to the results or observational data shown in a figure.

Coming back to our search for *universalities* in Biology, we find that, in contrast to the bewildering complexity and diversity of life, mentioned above, there is a remarkable linear dependence between the logarithms of the metabolic rate and body weight of many large and small animals (birds and mammals): see Fig. 12.4.[10] In other words the ratio k between the two logarithms scales isometrically. This relationship is known as Kleiber's Law, after the pioneering work in this field by Kleiber [11].

For a biologist it is, of course, of great interest to understand the reason for this result, its range of validity and the causes of the (small) deviations, which can be observed in Fig. 12.4 and, above all, the significance of the ratio k. For us, however, it is more relevant to observe that the metabolic rate must somehow be related to the growth rate of the animals. Hence we ask whether there also exists some scaling law for their growth vs. time.

It turns out that such a universal law does exist and has been found [12, 13] by a physicist, Geoffrey B. West,[11] and his colleagues: see Fig. 12.5. If we call m(t) the

---

[9]From greek: *iso* = same, and *allo* = different.

[10]This image has been made available under the Creative Commons Attribution-Share Alike 3.0 Unported license (https://commons.wikimedia.org/wiki/File:Kleiber1947.jpg).

[11]G.B. West has also been President of the Santa Fe Institute[2] from 2005 to 2009.

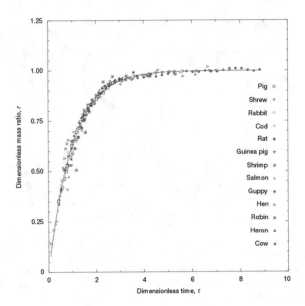

Fig. 12.5 A plot of the dimensionless mass ratio, $r = 1 - R = (m/M)^{1/4}$, versus the dimensionless time variable, $\tau = (at/4\ M^{1/4}) - \ln[1 - (m_0/M)^{1/4}]$, for a wide variety of determinate and indeterminate species. When plotted in this way, the model predicts that growth curves for all organisms should fall on the same universal parameterless curve $\sim 1 - e^{-\tau}$ (shown as a solid line). The model identifies r as the proportion of total lifetime metabolic power used for maintenance and other activities. Figure reprinted by permission from MacMillan Publishers Ltd: [Nature] (see Ref [12]), copyright (2001)

mass of a given animal at the time t after its birth, all the curves m(t) for a great variety of animals fall into a unique curve in which a dimensionless mass is plotted against a dimensionless time (both are defined in the figure caption). Since the curve is not linear, in this instance we have an allometric scaling law.

The striking beauty of the result shown in Fig. 12.5 impels the *Physicist within us* to search for other systems to which to apply similar procedures, in order to extract also from them other universal growth curves. And indeed at least two have been proposed so far: for the growth of cities [14] and of tumours [15]. Certainly many more *universalities* exist, waiting to be discovered, perhaps—who knows?—by some of our readers.

References

1. A. Tartaglia, Science and the Future Conference, Turin, Introduction (2013). arXiv:1312.0161 (Oct 28–31)
2. M. Waldrop, Complexity
3. P.P. Delsanto, *Universality of Nonclassical Nonlinearity* (Springer, 2007)

4. F.G. Tricomi, *Lezioni di Analisi Matematica* (1965)
5. G.B. West, J.H. Brown, Phys. Today **57**, 36 (2004)
6. P. Castorina, P.P. Delsanto, C. Guiot, Phys. Rev. Lett. **96**, 188701 (2006)
7. J.P. Freyer, R.M. Sutherland, Cancer Res. **46**, 3504 (1986)
8. P.P. Delsanto, C. Condat, N. Pugno, A.S. Gliozzi, M. Griffa, J. Theor. Biol. **250**, 16–24 (2008)
9. D. Mendeleev, *Principles of Chemistry* (in two volumes, 1868–1870)
10. G. Galilei, *Discorsi e Dimostrazioni Matematiche intorno à due nuove sci-enze attenenti alla Mecanica e i Movimenti Locali* (Discussions and mathematical demonstrations about two new sciences related to Mechanics and local movements, 1638)
11. M. Kleiber, Body size and metabolic rate. Physiol. Rev. **27**, 511–541 (1947)
12. G.B. West, J.H. Brown, B.J. Enquist, A general model for ontogenetic growth. Nature **413**, 628 (2001)
13. G.B. West, J.H. Brown, *Life's universal scaling laws* (Sept, Phys. Today, 2004), p. 36
14. L.M.A. Bettencourt, J. Lobo, D. Helbing, C. Kühnert, G.B. West, Growth, innovation, scaling, and the pace of life in cities. PNAS **104**(17), 7301 (2007)
15. C. Guiot, P.G. Degiorgis, P.P. Delsanto, P. Gabriele, T.S. Deisboeck, Does tumor growth follow a universal law? J. Theor. Biol. **225**, 147 (2003)

Chapter 13
Conclusions and Philosophical Implications

"The time has come," the Walrus said,
"To talk of many things:
Of shoes–and ships–and sealing-wax—
Of cabbages–and kings–
And why the sea is boiling hot–
And whether pigs have wings."

Carroll [1]

Abstract In the earlier chapters of this book, questions, discoveries and sometimes even paradoxes have been encountered which challenge or seem to challenge our intuition and our philosophical tenets. Without abandoning our path of a solid adherence to the scientific methodology required by physics, in this chapter we try to wrap up our conclusions over the current state-of-the art in physics and what it means for us human beings. Issues such as the Anthropic Principle, Variability of Physical Constants, Free Will versus Determinism, Quantum Entanglement, and the role of the observer in QM, together with known conflicts between Relativity and Quantum Mechanics, are discussed. Open questions are highlighted and (sometimes speculative) new ideas presented.

13.1 Introduction

If some readers, having now reached the final chapter in their journey through this book, have concluded that some of the concepts we have presented here are as likely as Lewis Carroll's flying pigs, then who could blame them. Every physicist has made the same voyage, and has only overcome his or her intuitive misgivings by the application of strict mathematical logic and the reliance on well-designed, careful experimentation. So has it been since the time of Galileo.

The potency of the scientific method is apparent from the highly technological world that we now live in. Our modern lifestyle would have been completely incomprehensible to those living only a century ago. And yet to give the impression that all is now understood, and that physics has finally reached the stage that Lord Kelvin believed had been attained at the end of the nineteenth century, i.e. where

© Springer International Publishing Switzerland 2016
R. Barrett et al., *Physics: The Ultimate Adventure*, Undergraduate Lecture Notes in Physics, DOI 10.1007/978-3-319-31691-8_13

195

everything important was known and only a few details remained to be filled in, is quite erroneous.

In this chapter we have gathered together a selection of those areas where open questions remain, or where we feel further discussion is warranted. Others would no doubt come up with a different choice. However, our aim is simply to highlight that physics is still an exciting field of study, and is likely to remain so for a very long time yet.

13.2 Anthropic Principle

In various places throughout this book, we have remarked that, unless certain *physical constants* have very specific values, the universe as we know it could not exist. Indeed, the conditions that would prevail in any universe with slightly different values of these parameters would preclude the development of intelligent life as we know it.

For example, as we have seen in Chap. 9, astronomer Fred Hoyle conjectured that the synthesis inside stars of elements beyond carbon requires the presence of an excited state of the ^{12}C nucleus at 7.6 MeV[1]; this state was subsequently discovered by Hoyle's colleague, Willie Fowler. Another example is the strength of the strong nuclear interaction. A slight increase (2 %) in the strength of this force would have enabled all the hydrogen in the early universe to bind into diprotons, rather than deuterium. This would have drastically altered the physics of stars, which are the ovens where the elements essential to life are forged.

Other examples of such *physical constants* are the velocity of light in a vacuum, Planck's constant and many others (some would say about 200). The idea that the universe is somehow fine-tuned to enable the development of intelligent life is known as the *Anthropic Principle*.

The Anthropic Principle comes in two forms: the strong and the weak. In the Weak Anthropic Principle, sentient beings find themselves in a time and place in the universe where the physical and cosmological constants take on values suitable for the evolution of life, and the universe is old enough for this to have already taken place. In the Strong Anthropic Principle the universe has somehow been compelled to evolve in a manner that favours the development of life. Some people believe that the latter invokes a divine purpose, and hence a creator. An alternative view is that the universe has undergone a form of *Darwinian selection* by the process of observation. If a universe were not capable of supporting sentient life, no

---

[1] 1 MeV = 1.602×10^{-13} Joules, and is the energy acquired by an electron when it is excited through a potential difference of 1 million volts.

observation of it could be made. Hence any universe that has been observed must have parameters tuned to the existence of intelligent life. Implicit is the idea that multiverses exist (see Chap. 8), and that we must live in one of these that supports life.

Perhaps these ideas can be made a little clearer with an allegory.

Once upon a time a beautiful butterfly emerged from its chrysalis and while she was stretching and drying her wings on a branch in the gentle sunlight, she looked up towards the sky, and lo, she beheld a spectacular arc of colour. This beautiful thing has been constructed solely for my delectation, she thought. In her period as a pupa in the chrysalis, she had immersed herself in the study of physics. She knew that light could be refracted by water drops, and she calculated that somehow water droplets of a particular size had been fixed in the sky at exactly the right locations to refract the rays of the sun back into her wonderstruck eyes. Drops of any other size located at any other angles would not produce this awe-inspiring display. She was a very privileged little lepidopterum, and felt uplifted that she alone of all the butterflies in the world should be selected for such a vision.

So excited was she that she waved a wing at a passing magpie,[2] who descended and regarded her with some surprise. He was unused to being hailed by what he regarded as a pre-dinner snack. He was even more nonplussed when she spoke to him and explained the special status she had been accorded. Now the magpie was a wise old bird who knew a thing or two about the way the world worked. He had just returned from a shower under the garden sprinkler, and was familiar with the rainbow that hung above it every sunny day. He tried to explain to the butterfly that she was nothing special, that there was a multitude of droplets passing through the sky from the sprinkler and that some of these were always going to be of the right size and in the right place to refract the light into her eyes. It was the same for everybody in the garden, not just her.

She didn't believe him of course. As he was cleaning the scales of the butterfly wings from his beak against the bark of an old gum tree, he gazed up at the rainbow. Yes, it is indeed a truly beautiful sight, he thought. He warbled and flew to a higher branch to spread his wings and enjoy a post-prandial snooze in the warm spring sunshine.

It is fair to say that the Anthropic Principle is highly contentious. Some eminent scientists, including Stephen Hawking and Freeman Dyson, maintain that it is a necessary evil to understand the many coincidences in the values of the physical constants. Others are reluctant to accept the Principle, and believe that future developments in physics will render it superfluous.

In Quantum Mechanics, the observer plays an important role. For instance, in a two-slit interference experiment with light, we cannot say where a photon is located, or which of the two slits it has passed through, until it has been observed (see later section in this chapter on Entanglement Revisited). Some physicists

---

[2]Scientific name: *Gymnorhina tibicen.*

extend this argument to the universe as a whole [2], and maintain that a universe without a sentient observer does not exist.[3]

13.3 Variability of the Physical Constants

Let us now return our attention to the physical constants, which play such a crucial role in the mathematical formulation of the physical laws, and their application to the natural world. Some of these constants eventually turn out to be derivable from others, and can be removed from the list of *fundamental* physical constants. It is assumed that the latter cannot be derived, and that their values must be determined by experiment.

The already mentioned physicist, George Gamow, wrote an entertaining book [3] on what the world would be like if the velocity of light were 10 miles/h, so that relativistic effects would become commonplace. Indeed there is no apparent reason, other than the one we discussed in the last section, why the velocity of light (or any other fundamental constant) has the value it has.

In Chap. 3 we saw the importance of defining our physical units of measurement carefully. However, all such definitions are arbitrary, and any physical constant expressed in such units must also be arbitrary. It is therefore customary to express fundamental constants in dimensionless units; for instance if we consider the ratio of the electron mass to the proton mass, we obtain a dimensionless number that is independent of any arbitrariness arising from the choice of units. Twenty-five fundamental dimensionless constants have now been identified whose numerical values are not understood in terms of any widely accepted theory, and must be determined from experiment [4].

Perhaps the best known of the fundamental constants is the fine structure constant α characterizing the strength of the electromagnetic interaction between elementary charged particles. Its value [5] is normally expressed as a reciprocal, i.e.

$$1/\alpha = 137.035999037(91)$$

with an incredibly small relative uncertainty of 6.6×10^{-10}. Richard Feynman, who won the Nobel Prize in Physics for his work on Quantum Electrodynamics, had this to say about the fine structure constant, and its apparent arbitrariness:

> *Immediately you would like to know where this number for a coupling comes from: is it related to pi or perhaps to the base of natural logarithms? Nobody knows. It's one of the greatest damn mysteries of physics: a magic number that comes to us with no*

---

[3]A question of semantics arises here on the definition of universe. *The* universe should be everything that exists. *A* universe instead (different from ours) refers to a portion of *the* universe that is outside all possibility of our observation. In any case, we should distinguish between the *visible* universe and *the* universe. As a matter of fact, that which cannot be observed (not even in principle) is outside the domain of physics.

understanding by man. You might say the "hand of God" wrote that number, and we don't know how He pushed his pencil. We know what kind of a dance to do experimentally to measure this number very accurately, but we don't know what kind of dance to do on the computer to make this number come out, without putting it in secretly! [6].

The very name of these quantities, i.e. *constants*, indicates that there is an implicit belief that their values are immutable throughout the past and future of the universe. Likewise there is a belief that the laws of physics are also immutable. As we have seen earlier, the laws of conservation of momentum and energy are found to be valid from sub-atomic to cosmological scales. However, the immutability of physical laws and constants is in reality a credo for which Occam's razor provides some justification—i.e., unchanging laws and constants provide a *simpler* scenario, that will continue to be accepted unless strong conflicting experimental evidence becomes available.

From time to time prominent physicists, such as Sir Arthur Eddington and Paul Dirac, father of Relativistic Quantum Mechanics, have floated the idea that some of the fundamental constants may change their values over time. The latter proposed that the value of the gravitational constant G decreases by 5 parts in 10^{11} per year as the universe expands. This rate of change is far too small to be detected by current terrestrial experiments. Philosopher, Alfred North Whitehead, opined that the laws of Physics themselves must change [7]:

Since the laws of nature depend on the individual characters of the things constituting nature, as the things change, then consequently the laws will change. Thus the modern evolutionary view of the physical universe should conceive of the laws of nature as evolving concurrently with the things constituting the environment. Thus the conception of the Universe as evolving subject to fixed eternal laws should be abandoned.

Queries have been raised from time to time about the constancy of the velocity of light. Over the period 1928–1945, the measured velocity of light fell steadily by an amount that was outside the bounds of quoted experimental error. Then in the late nineteen-forties new measurements were made that were in better agreement with the 1928 measurements. Had there really been a cyclic change in c over this time period? Most physicists believe not, and ascribe the result to a self-censorship process by experimenters. Measurements too far from the currently accepted value tend to be treated as statistical outliers, and discarded. The decision (see Chap. 3) to define the metre in terms of c means that what once would have been interpreted as a change in c would today appear as a change in the length of a metre. This is an example of why *dimensionless* constants are to be preferred when investigating the immutability of physical constants.

The long term variability of physical constants other than c has also been suggested, e.g. Planck's Constant, and the electronic charge. The question of whether the fine structure constant α is varying now, or has changed in the past, is an active research field. A very tight bound on current variations of α has been set at one part in 10^{17} per year [8]. This result however does not preclude past variations. Research by a group at the University of New South Wales has shown a variation in α of approximately one part in 10^5 spanning ~ 23–87 % of the age of the universe [9]. More recent work by this group claims that the changes in α are different in different directions in the universe [10]. Clearly these results have important implications for cosmological theories, and confirmation by other experimenters is awaited.

13.4   Determinism Versus Free Will

For thousands of years philosophers have debated the issue of free will (where individuals are free to make choices that affect their destiny) and determinism (where the future has been preordained by actions in the past since time immemorial). The ethics of our western society is largely based on free will. How can we punish criminals if they are not responsible for their actions?

On the other hand, determinism has been supported often by those to whom it provided an advantage, e.g. in arguments over the divine right of kings, the privileges of aristocracy, and the "elect" of Calvinism. A popular Anglican hymn, "All Things Bright and Beautiful", commonly sung by children, is indicative of this attitude. The words of the first stanza are harmless enough:

All things bright and beautiful,
All creatures great and small,
All things wise and wonderful,
The Lord God made them all.

However, the third stanza uses determinism to reinforce the class structure of the contemporary English society and is not so popular today:

The rich man in his castle,
The poor man at his gate,
God made them high and lowly,
And ordered their estate.

There are obvious limitations to free will. However much we may *will* ourselves to fly by flapping our arms like a bird's wings, it is not going to happen. The laws of physics preclude it. Hard-line determinists would maintain that everything is prescribed by physical laws, leaving no room for free will. Although we might think that we make *rational decisions* weighing up pros and cons, they would say that our decisions are already determined at our birth by the nature and state of the molecules of our bodies and of our world.

This idea was articulated by Pierre-Simon Laplace in 1814. He envisaged a "demon" who knew the location and momentum of all the particles in the universe. By applying the laws of classical physics he could calculate future and past values of these quantities. In fact, the history and future of the universe are completely determined because of the deterministic nature of Newton's classical mechanics.

But what is the situation after Einstein? As we have seen in Chap. 7, Einstein's theory of relativity leads also to a relativity of simultaneity, i.e. two events may appear simultaneous to one observer, but happening one after the other to a different observer. In other words, what is future depends on where we are and how we move. Thus we are led to believe that the future must be as real as the present and the past; i.e. everything must be already there. Then what about free will? One may have the illusion of building or changing the future, but it is not so. In Einstein's relativity, the world is already fully 'written' in the four-dimensional book of space-time.

It would then appear that physics comes down strongly on the side of determinism in the debate: determinism versus *free will*. However, even in the nineteenth century arguments were put forward against Laplace's demon. We have seen in Chap. 5 that statistical methods are necessary to explain whole branches of physics. The Second Law of Thermodynamics is statistical in nature, and some thermodynamical processes are deemed irreversible, leading to the concept of an Arrow of Time and an inexorable increase of entropy. If physical processes are irreversible, it is not possible to recover past situations from the current state. However, other interpretations are possible, in which statistical considerations and microscopic interactions are kept separate.

In QM the Heisenberg Uncertainty Principle does not allow a simultaneous precise observation of a particle's momentum and position, which are the prerequisite pieces of information needed by Laplace's demon. It might be argued that it is the process of observation itself which disturbs the particle and thus prevents these precise measurements. Before the measurement takes place, the position and momentum *are* precisely defined. This was Einstein's argument. However, Bell has shown that there are no "hidden variables", and that, until the measurement causes the collapse of the wave function (see Chap. 8), the momentum and position are truly undefined.

Chaos theory has been used as an argument against determinism. We have seen in Chap. 12 the "butterfly effect", whereby a butterfly fluttering its wings in the Amazon Rain Forest is responsible for a hurricane in another continent. However, Chaos Theory is deterministic. Very small changes in the input may produce vastly different outcomes, but *precisely* the same input will produce exactly the same output. A demon such as Laplace's is expected to know the input with an infinite precision, and so he can still predict the final outcome. What we have discussed above highlights the discrepancy between the deterministic theory of relativity and the statistical nature of quantum mechanics. Resolution of this dilemma is an actively pursued research field.

13.5 Entanglement Revisited

One of the issues that has puzzled physicists over the past centuries has been "action at a distance." Another name for it is "non-locality". In Newton's time it was generally believed that the only way an object, or piece of matter, could affect another was by direct material contact. As we have seen in Chap. 4, a criticism made of Newton's law of gravity by Descartes, Leibnitz and others was that the theory proposed a mysterious mutual attraction of objects across large distances. Newton himself rejected the idea of action at a distance, and attempted to account for gravity by the pressure of some intervening medium.[4] However, he was unsuccessful and did not publish anything about these investigations.

Newton's approach was to investigate the properties of forces between bodies, without attempting to explain how the forces are transmitted, an approach adopted later by Cavendish, Coulomb and others for the investigation of the properties of electric and magnetic forces.

The nature of the electromagnetic field has been touched on in Chap. 8. Suffice it to say here that every particle is an excitation of a field, analogous to a wave in the ocean. The field interacts with other particles and their associated fields. The fields vibrate against each other, giving the appearance of new particles appearing. In "field theory" the concept of "action at a distance" is replaced by the concept of action through a field, induced in the space between the objects by their very presence.

Quantum Field Theory is a very abstract topic that involves advanced mathematics, and must remain outside the scope of this book. In it the force between two particles is generated by the exchange of virtual particles between them. For the electromagnetic field, the exchanged particles are virtual photons. The exchange of virtual particles is a way to express the transfer of energy in a quantized manner. The mathematical terms describing the interaction mimic particle-like quanta, which however cannot be observed: they are a QM consistent way of doing calculations. The attribute of "virtual" is also due to the fact that, if interpreted as particles, these quanta can violate conservation laws and constraints, which real particles cannot infringe. Considering the real observable outcome of an experiment, all particles, including the virtual ones, are always combined in such a way that the conservation laws, the arrow of time and the causality relations are never violated.[5]

Many theoretical physicists have also tried, and continue to try, to describe the gravitational interaction in terms of the exchange of massless spin 2 quanta, called *gravitons* (see Chap. 10). However all attempts so far have failed and the theory per se is inconsistent. The inconsistency comes from the fact that Gravity is *not* a field like the others, sitting in a continuous flat background like the Minkowski space-time of special relativity. Gravity, at least in General Relativity, *is* a curved space-time on which the other fields may be defined. In other words, the quantization processes imply a continuous background. Quantizing space-time somehow

---

[4]"That gravity should be innate, inherent, and essential to matter, so that one body can act upon another at a distance, through a vacuum, without the mediation of anything else, by and through which their action and force may be conveyed from one to another, is to me so great an absurdity, that I believe no man who has in philosophical matters a competent faculty of thinking can ever fall into it." Original letter from Isaac Newton to Richard Bentley, 189. R.4.47, ff. 7–8, Trinity College Library, Cambridge, UK, Published online: http://www.newtonproject.sussex.ac.uk/view/texts/normalized/THEM00258 (accessed 23rd September, 2015).

[5]This is true in the reference frame chosen for the observer describing the interaction. The problem of the inconsistency between QM and relativity, when different observers are involved, remains for virtual as well as for real particles.

implies (at least implicitly) a continuous 'super space-time' against which the 'ordinary' space-time should be quantized.

Although Quantum Field Theory has perhaps provided an explanation of the mechanisms involved in action at a distance, quantum entanglement has brought with it another collection of oddities. As we have seen in Chap. 8, two particles emitted in a process in which the conservation laws constrain their total spin, (e.g. to be zero), are said to be "entangled". Detection of one of the two particles collapses that particle's wave function, so that its spin direction becomes precisely specified. As a result and simultaneously, in order to maintain at zero the total spin of the pair, the spin of the other particle also becomes precisely defined (opposite to the spin of the first one), no matter how far away the other particle is. Once again, "spooky" instantaneous action at a distance, a phrase coined by Einstein himself, emerges as a possibility.

Einstein's disparaging comment was meant as a "put down" of QM. In a paper with Podolski and Rosen [11], he reasoned that quantum mechanics must be "incomplete" and argued for the existence of "hidden variables" to remove the indeterminism associated with it (as mentioned in the last section). However, a theorem [12] by Irish physicist John Stewart Bell ruled out this possibility, showing that no physical theory of local hidden variables can ever reproduce all of the predictions of QM. Experimental confirmation of the validity of Bell's Theorem has increased over the past several decades.

Another issue with quantum entanglement is that it demonstrates further conflict between QM and the Theory of Relativity. The measurement of one entangled particle affects its partner *simultaneously*. However, as we have seen in Chap. 7, simultaneity in relativity depends on the state of motion of the observer. Two observers in different reference frames may have a different perception of which measurement occurs first.

One way to address the puzzle of entanglement was suggested by Bell himself. It invokes *superdeterminism*, where every attribute of the universe is pre-ordained. There is no free will, and entanglement requires no faster-than-light transmission of signals from one particle to the other because the measurement and its outcome are already writ large in the book of fate. Needless to say, this explanation finds little appeal in scientific circles.

The standard reading of QM, in terms of probability waves that satisfy the Schrödinger equation (see Chap. 8), is known as the "Copenhagen" interpretation because of its championing by Niels Bohr. An alternative "pilot wave theory" was proposed by Louis de Broglie in 1927. Although it found little favour at the time, it was resurrected by David Bohm in 1952, and today has a small number of followers. In this interpretation there exists a real physical wave, the "pilot wave", and an actual physical particle that is propelled by the wave, rather than being defined by it. In the two slit experiment, the pilot wave passes through the two slits, whereas the particle is drawn to places where the two wavefronts reinforce, rather than cancel out. In principle the pilot wave theory is deterministic. Debate between adherents of the two interpretations is ongoing.

Despite the theoretical and philosophical problems raised by entanglement, a number of possible applications are under development. One consists in the creation of ultra-precise clocks: adding entangled atoms to an atomic clock effectively increases its precision. Other applications are under study in cryptography and in the development of super-fast computers. Recently, entangled photons have been used to produce an image of an object with photons entering the camera that have undergone no contact with the object being photographed [13]. A pair of entangled photons are produced, one heading off to the object, the other proceeding towards the camera. The latter knows of its twin's life, and can be used to build up an image. The two photons may have different frequencies, so that a low-frequency photon can probe the object and the image can be produced by a higher frequency photon.

It is safe to predict that the future will see many more applications of quantum entanglement.

13.6   Reality and the Role of the Observer

If a tree falls in a forest, and no one is around to hear, does it make a sound?

A similar question is asked in an old scholastic limerick[6]:

There was a young man who said "God
Must think it exceedingly odd,
If He finds that the tree,
Continues to be
When there's no one about in the quad."

Dear Sir, your astonishment's odd;
I'm always about in the quad.
And that's why the tree
Continues to be,
Since observed by yours faithfully—God.

Questions of this type have been asked many times by philosophers, at least from the time of Berkeley (1710) onwards [14]. Albert Einstein was reputed to have asked Neils Bohr whether he believed that the moon existed when nobody was looking at it. Bohr replied that it would not matter, since no one could prove it, one way or the other. These types of questions may appear to be totally futile, but some developments in QM have rekindled interest in them.

The importance of the observer in QM is highlighted by the Delayed Choice Experiment, initially a *Gedankenexperiment* (thought experiment) proposed by John Archibald Wheeler. Consider the setup outlined in Fig. 13.1.

At the left of the figure is a laser emitting a beam of light, which passes through a single slit from which it is diffracted onto the traditional double slit. It arrives at a

---

[6]Quoted by Tom Kerns in his lecture over Bishop George Berkeley (2012).

Fig. 13.1 Experimental
arrangement illustrating the
Delayed Choice Experiment
proposed by Wheeler

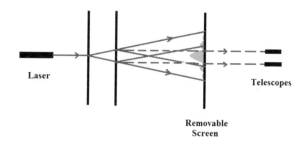

Laser   Telescopes

Removable
Screen

removable screen on which, over time, the distribution of photons builds up the
interference pattern first observed by Young (see Chap. 6). To the right of the
screen are two telescopes, which, when the screen is removed, are directed at the
two slits.

We thus have two separate methods of detecting the photons. With the first,
using the screen, we can measure the photon distribution across the screen, but we
cannot say which slit each photon passed through. With the screen removed, we
cannot observe the interference pattern, but we can determine which slit the photon
would have come through by noting which telescope sees a flash of light. After say
a hundred photons have passed through the slits, if we have the screen in place—
imagine it is a photographic plate—we will have obtained a picture of the distribution of photon arrivals across the screen revealing an interference pattern. Thus,
with the screen removed and using the two telescopes as detectors, we observe the
corpuscular behaviour of the photons, while with the screen in place we detect the
effect of their wave nature. It would appear that our choice as observers in deciding
which detection system to use affects the particle's "choice" on whether to behave
like a wave or a particle. The relevance of the question with which we began this
section now becomes clear.

However, there is one additional sting in the tail of this *Gedankenexperiment.*
Suppose we reduce the beam intensity so that only one photon is in flight at a time.
We then delay our choice on whether to use a screen or two telescopes for the
detection of each photon until it has passed through either one or both slits. The
photon will have made its "choice" on whether to be a particle (go through one slit
only) or a wave (go through both slits) at the time when it passes the slits, and now
it is too late to change. Right? *Wrong! The photon changes its mind, even when we
delay our choice so much that there is no time for a signal to go at speed c back to
the slits.*

One might like to dismiss Wheeler's *Gedankenexperiment* as a curious intellectual exercise that has no correspondence in the real world. However, there is
ample experimental evidence to show that what we have described above is exactly
what happens in reality [15]. Also, a group at the Australian National University has
recently carried out the Delayed Choice Experiment with a single atom [16]. Atoms
are relatively large particles to be involved in interference experiments of this type,
and the question arises of how large can an object be and still undergo these
quantum effects. The experiment confirms Niels Bohr's conjecture that the *choice*

between the wave or particle behaviour of a massive particle is determined by the measurement itself.

So the question remains: is the moon really there if no one is looking at it?

13.7   What Do We Really Know About the Universe?

After a full chapter devoted to Cosmology, the title of this section may look odd. However, despite having discussed both dark matter and dark energy, our observations extend only to the *visible* universe. Once we accept the idea of an origin of cosmic time at about 13.8 billion years in the past, we also recognize that no previous information may be available to us, or even have a meaningful existence, beyond this time. The distance travelled by light in such a time interval identifies a horizon beyond which we cannot know anything. And even that is a horizon only in principle, since our capacity to effectively obtain information is further hindered by the existence of another horizon not so distant in the past represented by the recombination era (see Chap. 11).

So a host of questions may be raised, almost with the certainty that they will remain unanswered. Consequently they belong more to the realm of speculation than of Physics. However, let us list a few of them here, just for the fun of it:

- Is space everywhere flat (as appears to be the case in the visible universe) and infinite? Could it be curved at a much higher scale than the one of the horizon, corresponding to an immense, but finite, universe?
- Is our expanding "universe" just a "local" inflating bubble starting from a singularity? Could other bubbles exist in other regions of the whole universe, expanding, collapsing or even exploding?[7] Exploring this possibility further, physicists sometimes use the plural "universes", considering these local "bubbles" to be sub-units of the global universe.
- As we mentioned before, the so called *constants of Physics* might not be constant in time. What if they are not constant in space, i.e. if in different areas of the global universe they have different values? Different "universes" would then exist with entirely different properties.
- Since we shall never have access to information from beyond the horizon, any speculation is possible with the only constraint being logical consistency. Of course, it is hard to accept these speculations as being part of physics, rather than metaphysics—unless QM has some surprise to bring us.
- Quantum fields have in principle an infinite extent, as do wave functions in free space. If, as evidenced by entanglement and despite its conflicts with relativity, a direct correlation exists between objects located *everywhere* in the universe, then what happens here every day is somehow influenced by the whole universe.

---

[7]The accelerated expansion scenario, if protracted into the future leads to what has been called *the big rip* when the expansion rate exceeds the speed of light.

This possibility opens the way to a holistic view of reality, in which our everyday reassuring physics recedes into the background.

- In a universe that is infinite and where the laws are the same as here, events that are extremely unlikely, such as a glass of water spontaneously boiling, become certainties. The expected repetition time for such an occurrence may be much longer than the age of the universe. However, if the universe is infinite with an infinite number of particles, unlikelihood is converted into virtual certainty: somewhere in the universe a glass of water is spontaneously boiling right now. Somewhere there are infinite copies of each of us, with all possible personal histories. We are reaching a scenario which looks like Borges's "Library of Babel": a library that contained all possible books, written in all possible languages and alphabets. When one plays with infinities, the possibilities are boundless.

13.8   Philosophical Implications of Relativity

The crucial variable of relativity is time, an old enigma pervading human thought. Already about 1600 years ago Saint Augustine was writing: "What is time? If nobody asks me, I know. If I must answer the question, I don't".[8] Time, we think, is something that *'flows equably without regard to anything external'*, as Newton wrote in a scholium of the *Philosophiae Naturalis Principia Mathematica*. But what does it mean?

Past no longer exists; future does not exist yet. So apparently only the present exists, but for a fleeting moment only. In other words, time is something that carries nothing to nothing. How can physicists use such a strange thing in their equations?

Einstein gave a simple explanation to the problem of time. He reduced it to an additional dimension of a four-dimensional continuum: the space-time. We have already discussed the deterministic nature of his theories, and the relativity of simultaneity that is a feature of them.

There are two more issues that it is worth mentioning in this section. One is the duality between space-time and matter/energy. As we have seen, these two ingredients of reality influence each other in a way prescribed by the Einstein equations. Apparently, however, matter/energy cannot be described or even thought about without space-time, whereas the latter can in principle stand alone. We can think of an empty and flat space-time, but we do not know what matter/energy could be out of space-time. Looking at his equations Einstein used to say that one side (the one containing tensorial objects related to the curvature of the manifold) was 'marble', the other side (the one containing matter/energy) was 'wooden'. In other words, on one side there was self-consistent, clean and elegant geometry; on the other,

---

[8]S. Augustine, Confessions, XI, 14, end of the 4th century A.D. Author's translation of *'Quid est tempus? Si nemo a me quaerat, scio. Si quaerenti explicare velim, nescio'*.

something empirically described without a full logical framework. Is the physical world really dual? Or is matter/energy an accident of space-time?

The final question concerns the reliability of some of the concepts we use in general relativity (in fact in any physical theory). A typical example is the massive point particle. A point is an abstract geometric entity, whose main property is to have no extension in space; in general relativity rather than a massive point particle, we should consider the corresponding world-line, extended in the time dimension. In our universe all real objects are of course extended objects, we could then be led to describe such objects, while in free fall, as bundles of geodetics of the given space-time.

The fact that the object is massive, however, poses a problem because any mass produces curvature around itself. In the Newtonian gravity it is possible to find exact analytical solutions to the two-body problem, provided the two bodies have a spherical mass distribution. The three body (or higher) interaction is treated approximately and perturbatively, but not analytically. In general relativity it is possible to solve exactly only the *one-body* problem, i.e. to determine the geometric structure of the space-time surrounding a given source, for instance in the case of the Schwarzschild symmetry. However it is impossible to have an analytical solution even for the two-body problem because the geometric structure of space-time is determined by both bodies and it is impossible to simply superpose the solutions valid for one body at a time. Currently one uses the one body solutions, then describes the free fall of other objects as happening along geodesics of those solutions, treating the falling object as a *test particle* i.e. a point particle with *negligible* mass.

An analogy might be a sphere rolling down a stretched elastic surface. If the sphere is lightweight, it will not distort the surface that it is rolling over. If however the sphere is heavy, it will stretch the elastic membrane, and its motion will be influenced by the consequent changes in the slope of the surface. In the one body solutions discussed in the preceding paragraph, the analogous distortions of space-time by a massive object are ignored.

Summing up, we may say that the properties of real objects, such as finiteness and extension, do not really fit into a purely geometric description of space-time.

13.9 Philosophical Implications of Quantum Mechanics

QM raises an even greater number of philosophical questions. We mention here some of the most relevant.

- As we have already seen, QM cannot predict the future, even if one knows all the initial data and laws. It can only determine the probability distribution of all possible futures. This difference between QM and classical physics, raises questions about the causality principle: apparently in QM there is no unique link among causes and effects, even though the fundamental physical constraints, such as the conservations laws, are always preserved.

- In most practical situations the gravitational interaction is de facto decoupled from all others (electromagnetic, nuclear and sub-nuclear forces), since at the atomic and nuclear scale, gravity is very much weaker than the other forces. The conflicts between relativity and QM that we have already discussed, are then less significant. There are, however, extreme situations (ultra-high energy and short distance interactions, such as are expected in the first instants after the Big Bang), in which gravity is comparable or even dominant with respect to the other forces. In these situations it is not possible to ignore that an entirely classical deterministic force would affect quantum objects and, above all, that quantum objects and their associated energies would be a relevant source of deterministic gravity.
  This inconsistency has led theoretical physicists to look for a way to *quantize gravity*. However, so far, nobody has succeeded. The deterministic nature of General Relativity is fundamentally incompatible with the probabilistic and statistical nature of QM. The arena of conflict is again in the concept of time.
- As we have seen, QM is abstract to the highest degree. It deals with *operators* and wave functions, state spaces and path integrals within those spaces. There is strong evidence for a non-physical nature of the wave function per se, even though it controls various physical effects. The ingredients of QM are essentially mathematical tools, and mathematics resides in our mind. What is then the relationship between reality and our mind? What is the nature of consciousness, and does QM play any part in it? These are questions, formerly regarded as in the domain of philosophy or metaphysics, that are now being investigated by eminent physicists [17].

13.10   Final Conclusions

We have already discussed several times the dilemma faced by physicists at the end of the nineteenth century, when two theories—classical mechanics and Maxwell's Electromagnetism—were found to be mutually contradictory. It took the genius of men like Max Planck, Albert Einstein, Erwin Schrödinger and Werner Heisenberg to resolve the issue. Their efforts gave us the Theory of Relativity, and QM, and astounded us with a picture of the world that lay far beyond anything our imaginations had hitherto conceived.

Now as we progress into the twenty-first century, we find ourselves in a similar situation. On closer inspection, the two theories that represent the pinnacles of the Golden Age of Physics appear themselves to be in contradiction. QM is a probabilistic theory, where only the statistical likelihood of an event can be calculated. Relativity is a deterministic theory, as is classical mechanics, where events can be foretold precisely. In fact, Relativity challenges the classical nature of time, with past, present and future depending on the perspective of the observer. Events which are simultaneous for one observer may not be simultaneous for another. And yet

QM tells us that if two particles are entangled and the wave function of one of them collapses after a measurement, then the wave function of the other also collapses *simultaneously*, no matter how far away it is. Simultaneous, but from whose point of view?

So where does physics go from here? Of course, we cannot say, but in this chapter we have presented some of the outstanding puzzles, and a few of the sometimes speculative ideas that have been put forward. Perhaps somewhere, labouring in a patent office in Berne, or commencing a Ph.D. course at one of the world's numerous universities, there is a new Einstein whose genius will resolve these enigmas. Or maybe many decades of slow and painstaking work by thousands of dedicated researchers will be necessary before any insight is gained. But all that lies in the future. Or is it the past?

References

1. L. Carroll, *The Walrus and The Carpenter* (from Through the Looking-Glass and What Alice Found There, 1872)
2. A. Linde, *Choose Your Own Universe*, ed. by C.H. Harper Jr., Spiritual In-formation (2005), p. 137
3. G. Gamow, *Mr Tompkins in Wonderland* (first published in 1940)
4. http://math.ucr.edu/home/baez/constants.html. Accessed 20 Dec 2015
5. R. Bouchendira, P. Cladé, S. Guellati-Khélifa, F. Nez, F. Biraben, Phys. Rev. Letts. **106** (2011). http://adsabs.harvard.edu/abs/2011PhRvL.106h0801B
6. R.P. Feynman, *QED: The Strange Theory of Light and Matter* (Princeton University Press, 1985), p. 129
7. Alfred North Whitehead, *Adventures of Ideas* (The Free Press, 1933)
8. T. Rosenband et al., Frequency Ratio of Al + and Hg + Single-Ion Optical Clocks; Metrology at the 17th Decimal Place. Science **319**(5871), 1808–12 (2008)
9. J.K. Webb, M.T. Murphy, V.V. Flambaum, V.A. Dzuba, J.D. Barrow, C.W. Churchill, J.X. Prochaska, A.M. Wolfe, Further evidence for cosmological evolution of the fine structure constant. Phys. Rev. Lett. **87**, 091301. (Published 9 August 2001)
10. J.K. Webb, J.A. King, M.T. Murphy, V.V. Flambaum, R.F. Carswell, M.B. Bainbridge, Indications of a spatial variation of the fine structure constant, Phys. Rev. Lett. **107**, 191101. (Published 31 Octo-ber 2011)
11. A. Einstein, B. Podolsky, N. Rosen, Can quantum-mechanical de-scription of physical reality be considered complete?. Phys. Rev. **47**, 777–780 (1935). (available online)
12. J.S. Bell, On the einstein-podolsky-rosen paradox. Physics **1**, 195–200 (1964)
13. G. Barreto Lemos, et al. Nature **512**, 409–412 (2014)
14. G. Berkeley, A treatise concerning the principles of human knowledge (1734). (section 23)
15. V. Jacques, E. Wu, F. Grosshans, F. Treussart, P. Grangier, A. Aspect, J.-F. Roch, Experimental re-alization of wheeler's delayed-choice gedanken experiment. Science **315** (5814), 966–968 (2007)
16. A.G. Manning, R.I. Khakimov, R.G. Dall, A.G. Truscott, Wheeler's delayed-choice Gedankenexperiment with a single atom. Nat. Phys. **11**, 539–542 (2015)
17. R. Penrose, *The Large, the Small, and the Human Mind* (Cam-bridge University Press, 1997)

Index

© Springer International Publishing Switzerland 2016
R. Barrett et al., *Physics: The Ultimate Adventure*, Undergraduate Lecture
Notes in Physics, DOI 10.1007/978-3-319-31691-8

Printed in the United States
By Bookmasters